Physical Science
and
Physical Science with Earth Science

Laboratory Activities Manual
Student Edition

McGraw Hill Education

Answers to the worksheets on pages vi-x can be found in the
Teacher Edition.

Credits
vi Matt Meadows Photography; vii & vii The McGraw-Hill Companies;
x & ix Courtesy Sargent-Welch/VWR Scientific Products.

The McGraw·Hill Companies

 Education

Send all inquiries to:
McGraw-Hill Education
8787 Orion Place
Columbus, OH 43240-4027

ISBN: 978-0-07-896281-3
MHID: 0-07-896281-1

Printed in the United States of America.

5 6 7 8 9 10 MAL 15 14 13 12 11

Table of Contents

Getting Started

Science is the body of information including all the hypotheses and experiments that tell us about our environment. All people involved in scientific work use similar methods for gaining information. One important scientific skill is the ability to obtain data directly from the environment. Observations must be based on what actually happens in the environment. Equally important is the ability to organize these data into a form from which valid conclusions can be drawn. These conclusions must be such that other scientists can achieve the same results in the laboratory.

To make the most of your laboratory experience, you need to continually work to increase your laboratory skills. These skills include the ability to recognize and use equipment properly and to measure and use SI units accurately. Safety also must be an ongoing concern. To help you get started in discovering many fascinating things about the world around you, the next few pages provide you with:

- a visual overview of basic **laboratory equipment** for you to label
- a reference sheet of **SI units**
- a reference sheet of **safety symbols**
- a list of your **safety responsibilities** in the laboratory
- a **safety contract**

Each lab activity in this manual includes the following sections:

- an investigation **title** and introductory section providing information about the problem under study
- a **strategy** section identifying the **objective(s)** of the activity
- a list of needed **materials**
- safety concerns identified with **safety icons** and **caution statements**
- a set of step-by-step **procedures**
- a section to help you record your **data and observations**
- a section to help you **analyze your data** and record your **conclusions**
- a closing **strategy check** so that you can review your achievement of the objectives of the activity

Laboratory Equipment

Figure 1

1. _____ 2. _____

3. _____

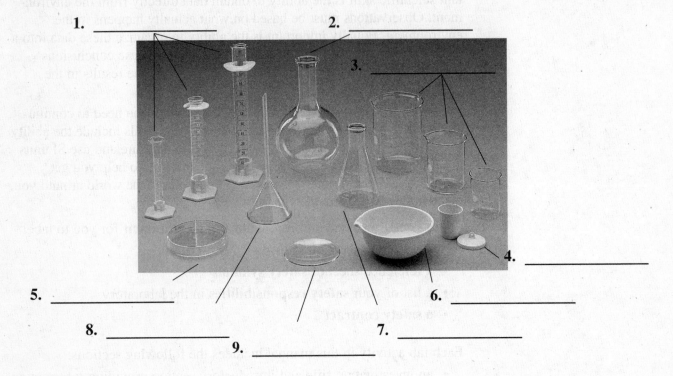

4. _____

5. _____ 6. _____

8. _____ 7. _____

9. _____

Figure 2

1. _____

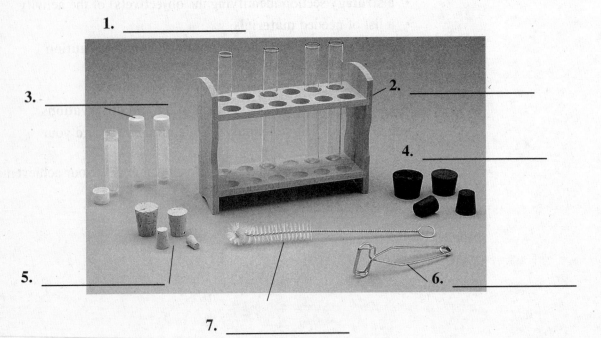

3. _____ 2. _____

4. _____

5. _____ 6. _____

7. _____

Laboratory Equipment (continued)

Figure 3

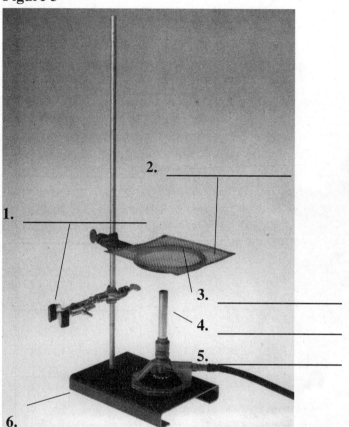

Figure 4

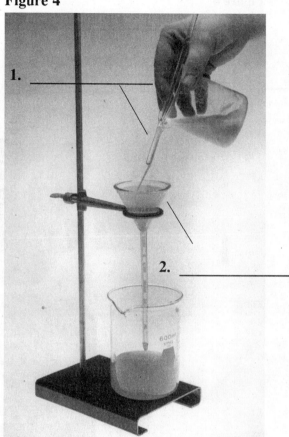

Figure 5

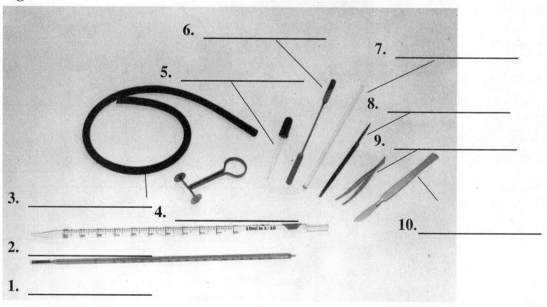

Laboratory Equipment (continued)

Figure 6

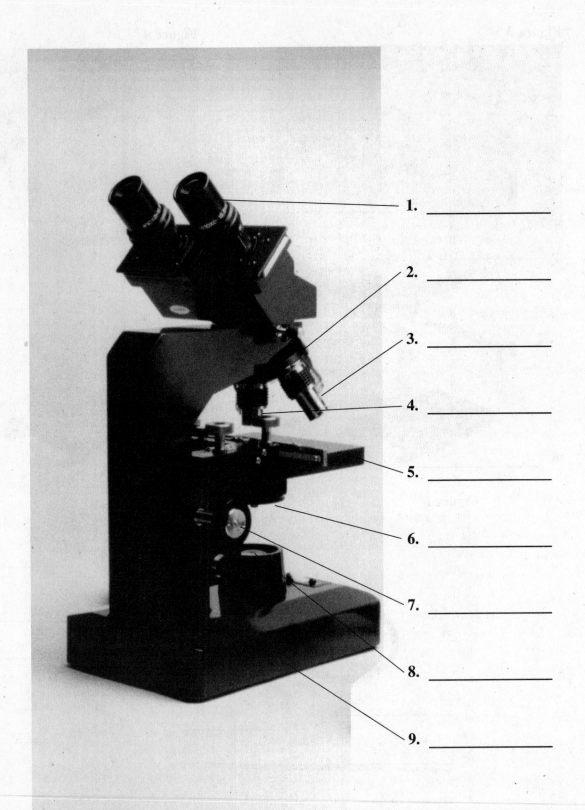

1. _____

2. _____

3. _____

4. _____

5. _____

6. _____

7. _____

8. _____

9. _____

Laboratory Equipment (continued)

Figure 7

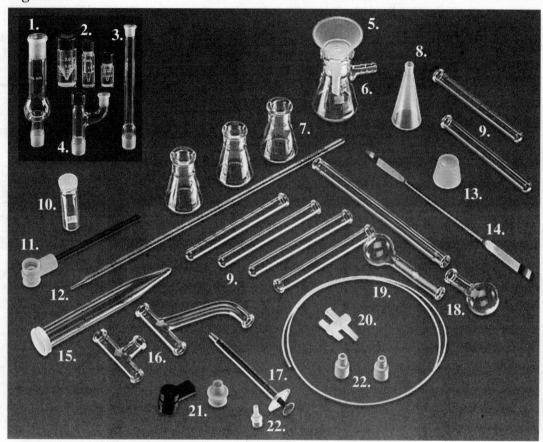

1. _____

2. _____

3. _____

4. _____

5. _____

6. _____

7. _____

8. _____

9. _____

10. _____

11. _____

12. _____

13. _____

14. _____

15. _____

16. _____

17. _____

18. _____

19. _____

20. _____

21. _____

22. _____

Laboratory Equipment (continued)

Figure 8

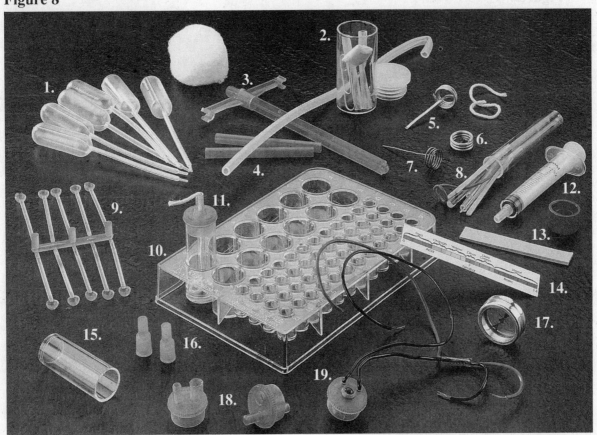

1. _____

2. _____

3. _____

4. _____

5. _____

6. _____

7. _____

8. _____

9. _____

10. _____

11. _____

12. _____

13. _____

14. _____

15. _____

16. _____

17. _____

18. _____

19. _____

SI Reference Sheet

The International System of Units (SI) is accepted as the standard for measurement throughout most of the world. Frequently used SI units are listed in **Table 1** and some supplementary SI units in **Table 2.**

Table 1

	Frequently Used SI Units
Length	1 millimeter (mm) = 100 micrometers (μm) 1 centimeter (cm) = 10 millimeters (mm) 1 meter (m) = 100 centimeters (cm) 1 kilometer (km) = 1,000 meters (m) 1 light-year = 9,460,000,000,000 kilometers (km)
Area	1 square meter (m^2) = 10,000 square centimeters (cm^2) 1 square kilometer (km^2) = 1,000,000 square meters (m^2)
Volume	1 milliliter (mL) = 1 cubic centimeter (cm^3) 1 liter (L) = 1,000 milliliters (mL)
Mass	1 gram (g) = 1,000 milligrams (mg) 1 kilogram (kg) = 1,000 grams (g) 1 metric ton = 1,000 kilograms (kg)
Time	1 s = 1 second

Table 2

Supplementary SI Units			
Measurement	**Unit**	**Symbol**	**Expressed in base units**
Energy	joule	J	$kg \cdot m^2/s^2$
Force	newton	N	$kg \cdot m/s^2$
Power	watt	W	$kg \cdot m^2/s^3$ or J/s
Pressure	pascal	Pa	$kg/m \cdot s^2$ or $N \cdot m$

Sometimes quantities are measured using different SI units. In order to use them together in an equation, you must convert all of the quantities into the same unit. To convert, you multiply by a conversion factor. A conversion factor is a ratio that is equal to one. Make a conversion factor by building a ratio of equivalent units. Place the new units in the numerator and the old units in the denominator. For example, to convert 1.255 L to mL, multiply 1.255 L by the appropriate ratio as follows:

$$1.255 \text{ L} \times 1,000 \text{ mL}/1 \text{ L} = 1,255 \text{ mL}$$

The unit L cancels just as if it were a number.

Temperature measurements in SI often are made in degrees Celsius. Celsius temperature is a supplementary unit derived from the base unit kelvin. The Celsius scale (°C) has 100 equal graduations between the freezing temperature (0°C) and the boiling temperature of water (100°C). The following relationship exists between the Celsius and kelvin temperature scales:

$$K = °C + 273$$

SI Reference Sheet (continued)

To convert from °F to °C, you can:

1. For exact amounts, use the equation at the bottom of **Table 3**, or
2. For approximate amounts, find °F on the thermometer at the left of **Figure 1** and determine °C on the thermometer at the right.

Table 3

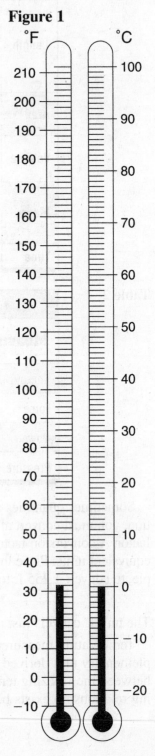

Figure 1

SI Metric to English Conversions			
	When you have:	**Multiply by:**	**To find:**
Length	inches	2.54	centimeters
	centimeters	0.39	inches
	feet	0.30	meters
	meters	3.28	feet
	yards	0.91	meters
	meters	1.09	yards
	miles	1.61	kilometers
	kilometers	0.62	miles
Mass and weight*	ounces	28.35	grams
	grams	0.04	ounces
	pounds	0.45	kilograms
	kilograms	2.20	pounds
	tons	0.91	metric tons
	metric tons	1.10	tons
	pounds	4.45	newtons
	newtons	0.23	pounds
Volume	cubic inches	16.39	cubic centimeters
	milliliters	0.06	cubic inches
	cubic feet	0.03	cubic meters
	cubic meters	35.31	cubic feet
	liters	1.06	quarts
	liters	0.26	gallons
	gallons	3.78	liters
Area	square inches	6.45	square centimeters
	square centimeters	0.16	square inches
	square feet	0.09	square meters
	square meters	10.76	square feet
	square miles	2.59	square kilometers
	square kilometers	0.39	square miles
	hectares	2.47	acres
	acres	0.40	hectares
Temperature	Fahrenheit	$\frac{5}{9}$ (°F − 32)	Celsius
	Celsius	$\frac{9}{5}$ °C + 32	Fahrenheit

* Weight as measured in standard Earth gravity

Copyright © Glencoe/McGraw-Hill, a division of The McGraw-Hill Companies, Inc.

Safety Symbols

These safety symbols are used in laboratory and investigations in this book to indicate possible hazards. Learn the meaning of each symbol and refer to this page often. *Remember to wash your hands thoroughly after completing lab procedures.*

SAFETY SYMBOLS	HAZARD	EXAMPLES	PRECAUTION	REMEDY
DISPOSAL	Special disposal procedures need to be followed.	certain chemicals, living organisms	Do not dispose of these materials in the sink or trash can.	Dispose of wastes as directed by your teacher.
BIOLOGICAL	Organisms or other biological materials that might be harmful to humans	bacteria, fungi, blood, unpreserved tissues, plant materials	Avoid skin contact with these materials. Wear mask or gloves.	Notify your teacher if you suspect contact with material. Wash hands thoroughly.
EXTREME TEMPERATURE	Objects that can burn skin by being too cold or too hot	boiling liquids, hot plates, dry ice, liquid nitrogen	Use proper protection when handling.	Go to your teacher for first aid.
SHARP OBJECT	Use of tools or glassware that can easily puncture or slice skin	razor blades, pins, scalpels, pointed tools, dissecting probes, broken glass	Practice common-sense behavior and follow guidelines for use of the tool.	Go to your teacher for first aid.
FUME	Possible danger to respiratory tract from fumes	ammonia, acetone, nail polish remover, heated sulfur, moth balls	Make sure there is good ventilation. Never smell fumes directly. Wear a mask.	Leave foul area and notify your teacher immediately.
ELECTRICAL	Possible danger from electrical shock or burn	improper grounding, liquid spills, short circuits, exposed wires	Double-check setup with teacher. Check condition of wires and apparatus.	Do not attempt to fix electrical problems. Notify your teacher immediately.
IRRITANT	Substances that can irritate the skin or mucous membranes of the respiratory tract	pollen, moth balls, steel wool, fiberglass, potassium permanganate	Wear dust mask and gloves. Practice extra care when handling these materials.	Go to your teacher for first aid.
CHEMICAL	Chemicals that can react with and destroy tissue and other materials	bleaches such as hydrogen peroxide; acids such as sulfuric acid, hydrochloric acid; bases such as ammonia, sodium hydroxide	Wear goggles, gloves, and an apron.	Immediately flush the affected area with water and notify your teacher.
TOXIC	Substance may be poisonous if touched, inhaled, or swallowed.	mercury, many metal compounds, iodine, poinsettia plant parts	Follow your teacher's instructions.	Always wash hands thoroughly after use. Go to your teacher for first aid.
FLAMMABLE	Open flame may ignite flammable chemicals, loose clothing, or hair.	alcohol, kerosene, potassium permanganate, hair, clothing	Avoid open flames and heat when using flammable chemicals.	Notify your teacher immediately. Use fire safety equipment if applicable.
OPEN FLAME	Open flame in use, may cause fire.	hair, clothing, paper, synthetic materials	Tie back hair and loose clothing. Follow teacher's instructions on lighting and extinguishing flames.	Always wash hands thoroughly after use. Go to your teacher for first aid.

 Eye Safety Proper eye protection should be worn at all times by anyone performing or observing science activities.

 Clothing Protection This symbol appears when substances could stain or burn clothing.

 Animal Safety This symbol appears when safety of animals and students must be ensured.

 Radioactivity This symbol appears when radioactive materials are used.

 Handwashing After the lab, wash hands with soap and water before removing goggles

Student Laboratory and Safety Guidelines

Regarding Emergencies

- Inform the teacher immediately of *any* mishap—fire, injury, glassware breakage, chemical spills, and so forth.
- Follow your teacher's instructions and your school's procedures in dealing with emergencies.

Regarding Your Person

- Do NOT wear clothing that is loose enough to catch on anything and avoid sandals or open-toed shoes.
- Wear protective safety gloves, goggles, and aprons as instructed.
- Always wear safety goggles (not glasses) when using hazardous chemicals.
- Wear goggles throughout entire activity, cleanup, and handwashing.
- Keep your hands away from your face while working in the laboratory.
- Remove synthetic fingernails before working in the lab (these are highly flammable).
- Do NOT use hair spray, mousse, or other flammable hair products just before or during laboratory work where an open flame is used (they can ignite easily).
- Tie back long hair and loose clothing to keep them away from flames and equipment.
- Remove loose jewelry—chains or bracelets—while doing lab work.
- NEVER eat or drink while in the lab or store food in lab equipment or the lab refrigerator.
- Do NOT inhale vapors or taste, touch, or smell any chemical or substance unless instructed to do so by your teacher.

Regarding Your Work

- Read all instructions before you begin a laboratory or field activity. Ask questions if you do not understand any part of the activity.
- Work ONLY on activities assigned by your teacher.
- Do NOT substitute other chemicals/substances for those listed in your activity.
- Do NOT begin any activity until directed to do so by your teacher.
- Do NOT handle any equipment without specific permission.
- Remain in your own work area unless given permission by your teacher to leave it.
- Do NOT point heated containers—test tubes, flasks, and so forth—at yourself or anyone else.
- Do NOT take any materials or chemicals out of the classroom.
- Stay out of storage areas unless you are instructed to be there and are supervised by your teacher.
- NEVER work alone in the laboratory.
- When using dissection equipment, always cut away from yourself and others. Cut downward, never stabbing at the object.
- Handle living organisms or preserved specimens only when authorized by your teacher.
- Always wear heavy gloves when handling animals. If you are bitten or stung, notify your teacher immediately.

Regarding Cleanup

- Keep work and lab areas clean, limiting the amount of easily ignitable materials.
- Turn off all burners and other equipment before leaving the lab.
- Carefully dispose of waste materials as instructed by your teacher.
- Wash your hands thoroughly with soap and warm water after each activity.

Teacher Approval Initials

Date of Approval

Student Lab Safety Form

Student Name: _____ Date: _____

Lab/Activity Title: _____

- Carefully read the entire lab and answer the following questions.
- Return this completed and signed safety form to your teacher to initial before you begin the lab/activity.

1. Describe what you will be doing during this lab/activity. Ask your teacher any questions you have regarding the lab/activity.

   _____

   _____

   _____

   _____

2. Will you be working alone, with a partner, or with a group? (Circle one.)

3. What safety precautions should you follow while doing this lab/activity?

   _____

   _____

   _____

4. Write any steps in the procedure, additional safety concerns, or lab safety symbols that you do not understand.

   _____

   _____

   _____

Student Signature _____

Relationships

LAB Laboratory Activity **1**

Most students will agree that the longer they study for tests, the higher they score. In other words, test grades seem to be related to the amount of time spent studying. If two variables are related, one variable depends on the other. One variable is called the independent variable; the other is called the dependent variable. If test grades and study time are related, what is the independent variable—the test grades or the time spent studying?

One of the most simple types of relationships is a linear relationship. In linear relationships, the change in the dependent variable caused by a change in the independent variable can be determined from a graph. In this experiment you will investigate how a graph can be used to describe the relationship between the stretch of a rubber band and the force stretching it.

Strategy

You will measure the effect of increasing forces on the length of a rubber band.
You will graph the results of the experiment.
You will interpret the graph.

Materials

ring stand
ring clamp
several heavy books
3 rubber bands, equal lengths, different widths
2 plastic-coated wire ties, 10 cm and 30 cm long
metric ruler
100-g, 200-g, and 500-g masses

Procedure

1. Set up the ring stand, ring clamp, and books as shown in Figure 1.
2. Choose the narrowest rubber band. Securely attach the rubber band to the ring clamp with the 10-cm plastic-coated wire tie.
3. Measure the width of the rubber band. Record this value in Table 1 in the Data and Observations section.
4. Measure the length of the rubber band as it hangs from the ring clamp. Record this value in Table 1 as 0 mass.
5. Attach the 100-g mass to the bottom of the rubber band with the second wire tie. Measure the length of the stretched rubber band. Record this value in Table 1.
6. Remove the mass and attach the 200-g mass to the bottom of the rubber band. Measure the length of the stretched rubber band. Record this value in Table 1.

Figure 1

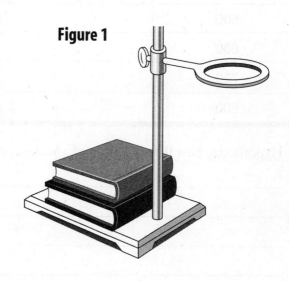

Laboratory Activity 1 (continued)

7. Remove the 200-g mass from the rubber band. Securely wrap the 100-g and 200-g masses together with the wire tie and tighten it. Attach the combined masses to the rubber band with the wire tie. Measure the length of the rubber band and record the value in Table 1.

8. Repeat measuring the lengths of the stretched rubber band for the 500-g mass and the combined masses of 600 g, 700 g, 800 g. Record the values in the data table.

9. Remove the rubber band.

10. Replace the rubber band with a slightly wider one. Hypothesize how the stretching of the wider rubber band will differ from that of the thinner one. Record your hypothesis in the Data and Observations section.

11. Repeat steps 3–9 for the second rubber band.

12. Replace the rubber band with the widest one and repeat steps 3–9 for the third rubber band.

Data and Observations

Table 1

Mass (g)	Length of rubber band (cm)		
	_____ mm width	_____ mm width	_____ mm width
0			
100			
200			
300			
500			
600			
700			
800			

1. Hypothesize how the stretching of a wider rubber band will differ from that of a thinner one.

_____

_____

_____

_____

_____

Laboratory Activity 1 (continued)

2. In most experiments, the independent variable is plotted on the *x*-axis, which is the horizontal axis. The dependent variable is plotted on the *y*-axis, which is the vertical axis. In this experiment, the lengths of the rubber bands change as more mass is used to stretch them. The length of each of the rubber bands is the dependent variable. The mass that is used to stretch them is the independent variable. Use Graph 1 to plot the data for all three rubber bands. Plot the values of the masses causing the rubber bands to stretch on the *x*-axis. Plot the lengths of the rubber bands on the *y*-axis. Label the *x*-axis *Mass (g)* and the *y*-axis *Length (cm)*.

Graph 1

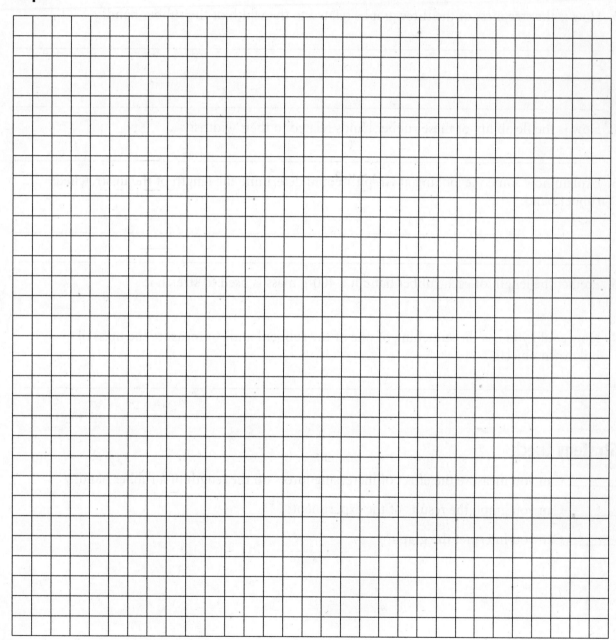

Laboratory Activity 1 (continued)

Questions and Conclusions

1. What do the graphs you made describe?

_____

_____

2. What does the steepness of the line of the graph measure?

_____

3. How is the steepness of the three graphs related to the width of the rubber band?

_____

_____

_____

4. How is the flexibility of these rubber bands related to their widths?

_____

5. Explain how someone looking at Graph 1 could determine the length of the unstretched rubber band.

_____

_____

6. Predict the length of each rubber band if a 400-g mass is used to stretch it.

_____

7. How could you use the stretching of one of the rubber bands to measure the mass of an unknown object?

_____

_____

Strategy Check

_____ Can you measure the effect of increasing forces on the length of a rubber band?

_____ Can you graph the results of the experiment?

_____ Can you interpret the graph?

No Need to Count Your Pennies

Have you ever saved pennies, nickels, or dimes? If you have, you probably took them to the bank in paper wrappers provided by the bank. Tellers at the bank could take the time to open each roll and count the coins to determine their dollar value. However, counting is not necessary because tellers use a better system. They use the properties of the coins instead.

A penny, a nickel, and a dime each has a particular mass and thickness. Therefore, a roll of coins will have a certain mass and length. These two properties—mass and length of a roll of coins—are often used to determine the dollar value of the coins in the roll.

Strategy

You will develop measuring skills using a balance and a metric ruler.
You will use graphing skills to make interpretations about your data.
You will compare the relationships among the mass, length, and number of coins in a roll.

Materials

10 coins (all of the same type)
balance
metric ruler
roll of coins

Procedure

1. Using the balance, determine the mass of 1 coin, 2 coins, 3 coins, 4 coins, 6 coins, 8 coins, and 10 coins to the nearest 0.1 g. Record the masses in Table 1 in the Data and Observations section.

2. Measure the thickness of 1 coin, 2 coins, 3 coins, 4 coins, 6 coins, 8 coins, and 10 coins to the nearest 0.5 mm. See Figure 1. Record these values in the table.

3. Record the number of coins in the roll on the table. Use the balance to find the mass of the roll of coins. Measure the length of the roll. Record these values in the table.

Figure 1

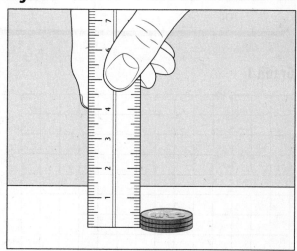

Laboratory Activity 2 (continued)

Data and Observations

1. Make two graphs of the information in Table 1. On Graph 1, show the number of coins on the x-axis and the mass of the coins on the y-axis. Graph 2 should compare the number of coins (x-axis) to the total thickness of the stacked coins (y-axis). Be sure to label each axis.

2. Draw a line connecting the points on each graph.

Table 1

Number of coins	Mass (g)	Thickness (mm)
1	3.1	2
2	5.6	3
3	8.1	4
4	11.1	6
6	14.2 16.7	9
8	21.6	11
10	27.2	15
roll =	134.5	75

Graph 1

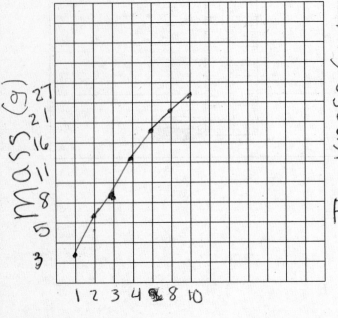

# of coins

Graph 2

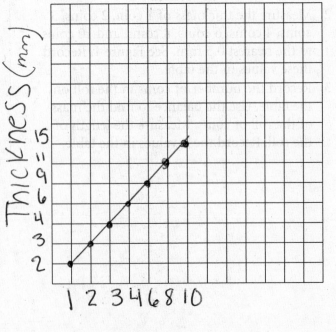

# of coins

Laboratory Activity 2 (continued)

(10)

Questions and Conclusions

1. Describe the appearance of the curve or line in each graph.

   They were mostly straight

2. What errors could exist in your measurement of the mass and the length of the coin roll?

   Depending on what you round it to your scale could be off

3. Which of the errors in question 2 would have real importance for a bank teller?

   Finding the exact amount of money in the roll

4. Do your data show a difference in the mass of different coins? Explain your answer.

   yeah, each coin weighed something different

5. Do your data show a difference in the thickness of different coins? Explain your answer.

   yeah, depending on what you round it to.

6. Could you use the mass of 1 coin to determine the mass of 2, 3, 4, 6, 8, and 10 coins? Why or why not?

   no, because each could weigh something different.

Strategy Check

_✓_ Can you develop measuring skills using a balance and a metric ruler?

_✓_ Can you use graphing skills to make interpretations about your data?

_✓_ Can you compare the relationships among the mass, length, and number of coins in a roll?

Projectile Motion

What do a volleyball, baseball, tennis ball, soccer ball, and football have in common? Each is used in a sport and each is a projectile after it is tapped, thrown, kicked, or hit. A projectile is any object that is thrown or shot into the air. If air resistance is ignored, the only force acting on a projectile is the force of gravity.

The path followed by a projectile is called a trajectory. Figure 1a shows the shape of the trajectory of a toy rocket. Because the force of gravity is the only force acting on it, the toy rocket has an acceleration of 9.80 m/s^2 downward. However, the motion of the projectile is upward and then downward. Figure 1b shows the size and direction of the vertical velocity of a toy rocket at different moments along its trajectory. The rocket's velocity upward begins to decrease immediately after launch and the rocket begins to slow down. The rocket continues to slow down. And then, for an instant at the highest point of its trajectory, it stops moving because its velocity upward is zero. As the rocket begins to fall, its velocity begins to increase downward.

As you can see, the shape of the upward trajectory of the rocket is a mirror-image of the shape of its downward trajectory. Can the trajectory of a toy rocket be used to learn something about the motion of a projectile? In this experiment you will find out.

Strategy

You will measure the flight times of a projectile.
You will analyze the flight times of a projectile.

Materials 🥽

toy water rocket and launcher
bucket of water
3 stopwatches

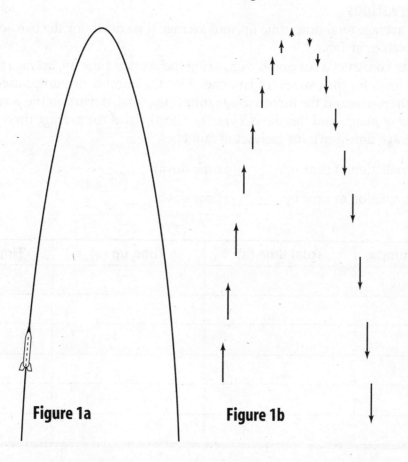

Figure 1a   **Figure 1b**

Laboratory Activity 1 (continued)

Procedure

1. Wear goggles during this experiment.
2. Fill the water rocket to the level line shown on the rocket's body. Always fill the rocket to the same level during each flight in the experiment.
3. Attach the pump/launcher to the rocket as shown in the manufacturer's directions.
4. Pump the pump/launcher 10 times. **CAUTION:** *Do not exceed 20 pumps or the maximum number suggested by the manufacturer, whichever is lower. Be sure to hold the rocket and pump/launcher so that the rocket is not directed toward yourself or another person.*
5. Launch the rocket vertically. Predict the time for the rocket to rise to its highest point, and the time for it to fall back to Earth. Now predict these times if the rocket is pumped 15 times. Record your predictions as time up and time down in the Data and Observations section.
6. Retrieve the rocket. Fill the rocket with water as in step 2. Pump the pump/launcher 10 times. Record the number of pumps in Table 1.
7. At a given signal to the timers, launch the rocket. Your teacher will have timers measure specific parts of the flight using stopwatches. Record the values measured by the timers as total time, time up, and time down in Table 1.
8. Repeat steps 6 and 7 twice.
9. Repeat steps 6 and 7 three more times, increasing the number of pumps to 15 for each launch. **CAUTION:** *Do not exceed the maximum number of pumps suggested by the manufacturer.*

Data and Observations

1. Calculate the average total time, time up, and average time down for the two sets of launches. Record these values in Table 2.
2. Use Graph 1 to construct a bar graph comparing the average time up, average total time, and average time down for the two sets of launches. Plot the number of pumps used in each set of launches on the *x*-axis and the three average times (up, total, down) on the *y*-axis. Label the *x*-axis *Number of pumps* and the *y*-axis *Time (s)*. Clearly label the average time up, average total time, and average time down for each set of launches.

10 pumps—Prediction of time up: _____; time down _____

15 pumps—Prediction of time up: _____; time down _____

Table 1

Number of pumps	Total time (s)	Time up (s)	Time down (s)

Laboratory Activity 1 (continued)

Table 2

Number of pumps	Average total time (s)	Average time up (s)	Average time down (s)

Graph 1

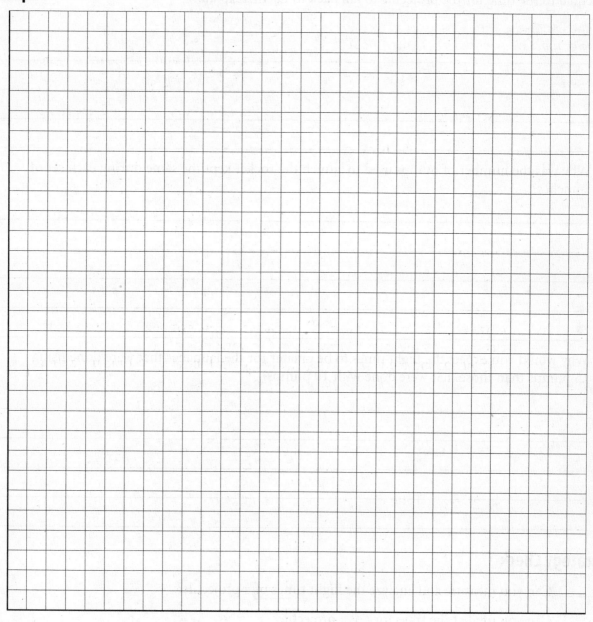

Laboratory Activity 1 (continued)

Questions and Conclusions

1. How well did your predictions agree with the measured times?

_____

_____

_____

2. Do your graphs support the statement that the time for a projectile to reach its highest point is equal to the time for the projectile to fall back to Earth? Explain.

_____

_____

_____

_____

_____

3. Why was the number of pumps used to launch the rocket kept the same during each set of launches?

_____

_____

_____

_____

_____

4. Why would you expect the flight times to be greater for the launches that were done using 15 pumps than those that were done with 10 pumps?

_____

_____

_____

_____

_____

Strategy Check

_____ Can you measure and analyze the flight times of a projectile?

_____ Can you predict the trajectory of a projectile?

Velocity and Momentum

Chapter 2

As you know, you can increase the speed of a shopping cart by pushing harder on its handles. You can also increase its speed by pushing on the handles for a longer time. Both ways will increase the momentum of the cart. How is the momentum of an object related to the time that a force acts on it? In this experiment, you will investigate that question.

Strategy

You will observe the effect of a net force on a cart.
You will measure the velocity of the cart at various times.
You will determine the momentum of the cart.
You will relate the momentum of the cart and the time during which the force acted on it.

Materials

utility clamp
ring stand
2 plastic-coated wire ties (1 short, 1 long)
pulley
metric balance
momentum cart
2–3 rubber bands
1-m length of string

100-g mass
3–4 books
plastic foam sheet
meterstick
masking tape
stopwatch/timer
felt-tip marker

Procedure

1. Attach the utility clamp to the ring stand. Using the short plastic-coated wire tie, attach the pulley to the clamp.

2. Use the metric balance to find the mass of the cart. Record this value in the Data and Observations section on the line provided.

3. Wrap the rubber bands around the cart lengthwise.

4. Tie one end of the string around the rubber bands as shown in Figure 1. Tie a loop at the opposite end of the string. Pass the string over the pulley.

5. Wrap the long plastic-coated wire tie securely around the 100-g mass. Attach the mass to the loop on the string with the wire tie.

6. Place the ring stand near the edge of the table. Adjust the position of the pulley so that the string is parallel to the table top as shown in Figure 2. Be sure that the 100-g mass can fall freely to the floor. Place several heavy books on the base of the ring stand.

7. Place a plastic foam sheet beneath the mass.

Figure 1

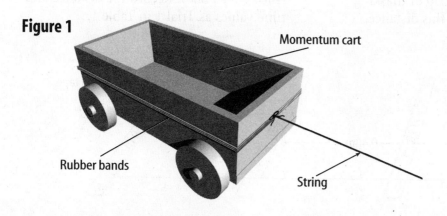

Momentum cart

Rubber bands

String

Laboratory Activity 2 (continued)

Figure 2

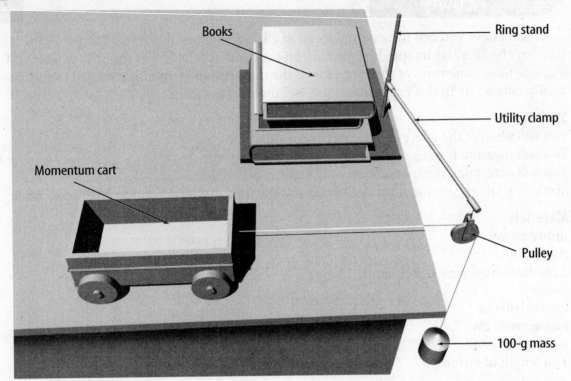

Books

Ring stand

Utility clamp

Momentum cart

Pulley

100-g mass

8. Pull the cart back until the 100-g mass is about 80 cm above the foam sheet. Have your lab partner place a strip of masking tape on the table marking the position of the front wheels. Release the cart. Observe the motion of the cart. Record your observations on a separate sheet of paper. **CAUTION:** *Have your partner stop the cart before it runs into the pulley.*

9. Using the marker, label the strip of masking tape *Starting Line.* Use the meterstick to measure a distance of 0.20 m from the starting line. Place a strip of masking tape on the table to mark this distance.

Be sure to have the strip of masking tape parallel to the starting line. Label the strip of masking tape *0.20 m.* Measure and label distances of 0.40 m and 0.60 m in the same manner. See Figure 3.

10. Pull the cart back with one hand until its front wheels are on the starting line. Hold the stopwatch in the other hand. Release the cart and immediately start the stopwatch. Measure the time for the front wheels to cross the 0.20-m line. **CAUTION:** *Have your partner stop the cart before it reaches the pulley.* Record the distance and time values as Trial 1 in Table 1.

Figure 3

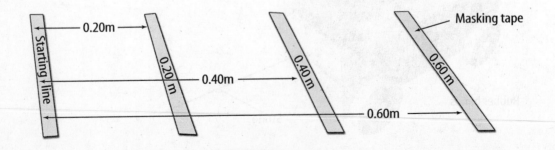

Masking tape

Starting line

0.20m

0.20 m

0.40m

0.40 m

0.60m

0.60 m

Laboratory Activity 2 (continued)

11. Repeat step 10 twice. Record the values as Trials 2 and 3.
12. Repeat steps 10 and 11 to measure the time for the front wheels to cross the 0.40-m and 0.60-m lines.

Data and Observations

Table 1

Distance (m)	Time (s)		
	Trial 1	Trial 2	Trial 3

Table 2

Distance (m)	Average time (s)	Average velocity (m/s)	Final velocity (m/s)	Momentum (g·m/s)

Graph 1

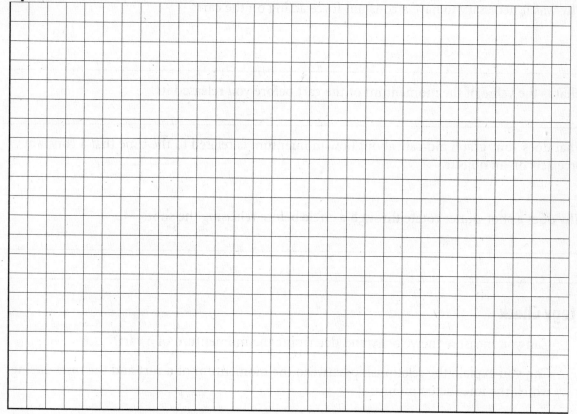

Laboratory Activity 2 (continued)

1. Calculate the average times for the cart to travel 0.20 m, 0.40 m, and 0.60 m. Record these values in Table 2.
2. Calculate the average velocity for each distance by dividing distance traveled by average time. Record these values in Table 2.
3. Because the cart started from rest and had a constant force acting on it, the velocity of the cart at a given distance from the starting line is equal to twice its average velocity for that distance. That is, the velocity of the cart as it crossed the 0.20-m line is twice the value of the average velocity that you calculated for 0.20 m. Calculate the velocity of the cart as it crossed the 0.20-m line, the 0.40-m line, and the 0.60-m line. Record these values in Table 2.
4. Calculate the momentum of the cart as it crossed the 0.20-m, 0.40-m, and 0.60-m lines by multiplying the mass of the cart by its velocity. Record these values in Table 2.
5. Use Graph 1 to make a graph of your data. Plot the average time on the *x*-axis and the momentum on the *y*-axis. Label the *x*-axis *Time (s)* and the *y*-axis *Momentum* (P).

Step 1. Mass of cart: _____ g

Step 8. Observation of motion of a cart: _____

_____

Questions and Conclusions

1. What force caused the cart to accelerate?

_____

2. Why was it necessary to have a constant force acting on the cart?

_____

_____

3. What is the value of the momentum of the cart before you released it?

_____

4. What does your graph indicate about how momentum is related to the time that a constant force acts on an object?

_____

5. Why does a shot-putter rotate through a circle before releasing the shot?

_____

_____

Strategy Check

_____ Can you measure the velocity and determine the momentum of a cart?

_____ Can you explain the relationship between the momentum of a cart and the time during which the force acted on it?

Pushing People Around

When we push something, we unconsciously compensate for how much mass it has. We know that if an object has a larger mass it will require more force to get it moving and if it has a small mass it will require less force. But how much difference is there? In this experiment, we will see what variables affect acceleration.

Strategy

You will see what happens when you use a constant force to pull a skater.
You will examine the relationship between force, acceleration, and mass.

Materials

tape
meter stick
roller skates
skating safety equipment (helmet, pads)
spring balance
stopwatch

Procedure

1. Mark positions on the floor at intervals of 0 m, 5 m, 10 m, and 15 m with the tape. The floor should be smooth, straight, and level.

2. Have one student stand on the 0-m mark with the skates on. A second student stands behind the mark and holds the skater. The skater holds the spring balance by its hook.

3. The third student holds the other end of the spring balance and exerts a constant pulling force on the skater. When the skater is released, the puller must maintain a constant force throughout the distance. Measure the time to each of the marks. Record this in the Data and Observations section along with the spring balance readings at each mark.

4. Repeat steps 2 and 3 for two different skaters in order to vary the mass. Keep the force the same. Make sure the skaters hold their skates parallel and do not try to change direction during the trial.

5. Repeat steps 2, 3, and 4 with a different constant force. Use the same three skaters. Record these results in the Data and Observations section.

Laboratory Activity 1 (continued)

Data and Observations

Table 1

	Roller Skater Distance, Trial 1		
Trial	Distance (m)	Force (N)	Time (s)
1	5		
	10		
	15		
2	5		
	10		
	15		
3	5		
	10		
	15		

Table 2

	Roller Skater Distance, Trial 2		
Trial	Distance (m)	Force (N)	Time (s)
1	5		
	10		
	15		
2	5		
	10		
	15		
3	5		
	10		
	15		

Laboratory Activity 1 (continued)

Questions and Conclusions

1. Until the time of Galileo and Newton, people believed that, disregarding friction, a constant force was required to produce a constant speed. Do your observations confirm or reject this notion?

_____

_____

2. What happens to the speed as you proceed farther along the measured distance?

_____

3. What happens to the rate of increase in speed—the acceleration—as you proceed farther along the measured distance?

_____

4. When the force is the same, how does the acceleration depend upon the mass?

_____

5. When the mass of the skater is the same, how does the acceleration depend upon the force?

_____

6. Suppose a 4-N force is applied to the skater and no movement results. How can this be explained?

_____

_____

Strategy Check

_____ Can you pull someone with a constant force?

_____ Can you explain the relationship between force, mass, and acceleration?

Friction Predictions

Friction can either be your friend or your foe. In some situations, friction might work in your favor, such as when the person driving the car you are riding in applies the brakes. At other times, you may want to do away with as much friction as possible, like when you are trying to open a stuck window. Different types of surfaces create different levels of friction and require different levels of force to move objects along them. This experiment will examine friction and show you the effect that different surfaces have on the total forces needed to move objects.

Strategy

You will observe and compare the forces needed to move an object over different types of surfaces. You will examine the principles of static, sliding, and rolling friction.

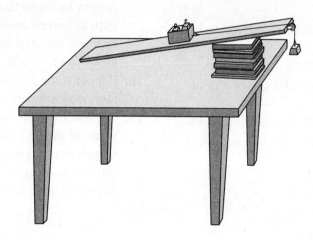

Materials

straight, smooth wooden board about
  1-meter-long
small pulley that can be screwed into the board
string
small open wooden box with hook on one end
stack of books
rough-grade sandpaper sheets

masking tape
vegetable oil
paper towels
50-g weights with hooks and a variety of other
  weights with hooks (8)
other surface materials that can be tested
  include strips of carpet, nylon material,
  adding machine paper, burlap, terry cloth,
  rubber or plastic.

Laboratory Activity 2 (continued)

Procedure

1. Attach the pulley to the end of the board so that the wheel is perpendicular to the length of the board.
2. Stack the books at the edge of the table.
3. Place the board on the books with the pulley end hanging over the table making sure that the pulley clears the edge of the table.
4. Attach one end of the string to the hook on the wooden box.
5. Place the box on the board and feed the string through the pulley so it hangs down the side of the table.
6. Make a loop at the end of the string to be able to hook the weights.
7. Place the weights in the box and at the bottom of the string so that the box does not move
8. Record the total weight on your data table.
9. Gradually add weight to the box so that it slides down the board

10. Record the total weight needed to move the box on your data table.
11. Add additional weight on the bottom of the string until the box begins to move up the incline.
12. Record the total weight to move the box.
13. Remove the weights from the box and-string and remove the box from the board.
14. Tape sandpaper to the board, making sure it is as smooth as possible where to sheets meet.
15. Predict if you will need more or less weight to move the box on the ramp using your different surfaces. Predict which surface will require the most and least amount of weight to move.
16. Repeat steps 5 through 13 for each surface, making sure each surface is taut.
17. To test an oiled board, pour some oil on a piece of paper towel and rub into the board, making sure the board is covered, but not so much that the oil is dripping. Repeat steps 5 through 13.

Data and Observations

Surface	Weight Needed to Keep the Box Still	Weight Needed to Move Box Downward	Weight Needed to Move Box Upward	Predictions	Observations
Wood					
Sandpaper					
Oiled Wood					
Other					
Other					
Other					

Laboratory Activity 2 (continued)

Questions and Conclusions

1. What data represents the static friction force in your experiment? Explain your answer.

   _____

   _____

   _____

2. What data represents the sliding friction force in your experiment? Explain your answer.

   _____

   _____

   _____

3. In which direction is the frictional force when the box slides down the plane? Explain your answer.

   _____

   _____

   _____

4. In which direction is the frictional force when the box slides upward on the plane?

   _____

5. What part of the apparatus demonstrates rolling friction?

   _____

6. Name three ways to make friction work in your favor.

   _____

7. Name three situations in which you would want to reduce friction.

   _____

8. Name three situations where rolling friction would be more advantageous than sliding friction.

   _____

Laboratory Activity 2 (continued)

Static and Sliding Friction Forces
for Different Surfaces

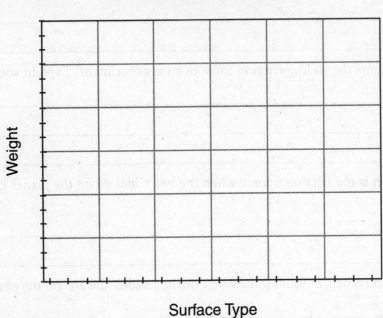

Weight

Surface Type

9. Make a bar graph comparing the static and sliding friction forces needed to move the box over each type of surface.

10. How do your predictions match your data?

_____

_____

_____

Strategy Check

_____ Can you explain the various types of friction?

Pulleys

If you have ever raised or lowered a flag or slatted blinds, you used a simple machine called a pulley. As you recall, simple machines can change direction of a force and multiply either the size of the effort force or the distance that the resistance force moves.

A single fixed pulley is a pulley that can't move up and down. As you can see in Figure 1, a fixed pulley is actually a lever in the form of a circle. Can you locate the effort arm and the resistance arm in a single fixed pulley?

Figure 1

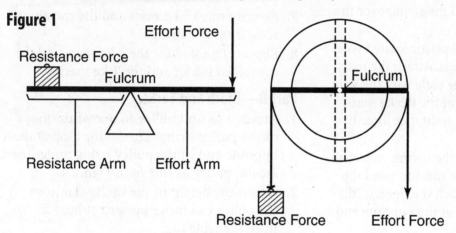

A series of pulleys is called a block and tackle. You may have seen a block and tackle in an auto repair shop. It sometimes is used to lift car engines. Look at the block and tackle shown in Figure 2. Can you locate a single fixed pulley in the block and tackle?

Figure 2

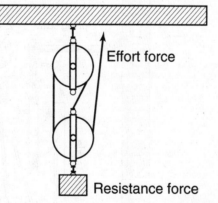

Strategy

You will perform work using a single fixed pulley.

You will construct a block and tackle and perform work with it.

You will compare the properties of a single fixed pulley and a block and tackle.

Materials

utility clamp
ring stand
plastic-coated wire ties, 10 cm and 30 cm long
2 pulleys
meterstick
1-m length of cotton string
masking tape
metric spring scale
0.5-kg and 1-kg standard masses

Laboratory Activity 1 (continued)

Procedure

Part A—Single Fixed Pulley

1. Attach the utility clamp to the top of a ring stand. Use the short plastic-coated wire tie to attach one of the pulleys to the utility clamp. Attach a meterstick to the ring stand with tape. See Figure 3.

2. Tie a small loop at each end of the 1-m length of string. Thread the string over the pulley.

3. Tightly wrap the second plastic-coated wire tie around the 0.5 kg mass. Attach the mass to the hook of the spring scale with the wire tie. Measure the weight of the 0.5 kg mass. Record this value as the resistance force in Table 1.

4. Remove the mass from the spring scale. Use the wire tie to attach the mass to one loop of the pulley string. Attach the hook of the spring scale to the loop at the opposite end of the string.

5. Slowly pull straight down on the spring scale to raise the mass. Measure the force needed to raise the mass 15 cm. Record this value as the effort force in Table 1.

6. Lower the mass to the table top. As you again pull down on the spring scale, measure the distance the spring scale moves as you raise the mass a distance of 15 cm. Record this value as the effort distance in Table 1.

7. Remove the 0.5 kg mass and the spring scale from the string.

8. Repeat steps 4–7 for the 1 kg mass and the combined 0.5 kg and the 1 kg masses.

Part B—Block and Tackle

1. Attach a second pulley to one of the loops of the pulley string. Thread the loop at the opposite end of the pulley string under the second pulley as shown in Figure 4.

2. Adjust the height of the utility clamp so the pulley can move upward at least 25 cm from the table top.

Figure 3

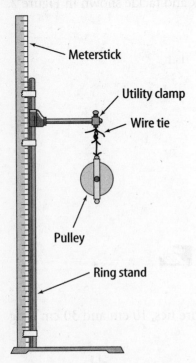

Figure 4

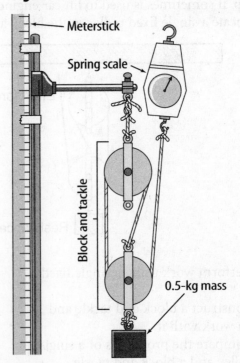

Laboratory Activity 1 (continued)

3. Wrap the plastic wire tie securely around the 0.5 kg mass. Use the spring scale to measure its weight. Record this value as the resistance force in Table 2. Attach the mass to the second pulley.

4. Attach the spring scale to the loop on the free end of the string.

5. Slowly pull straight up on the spring scale to raise the mass as shown in Figure 4. Measure the force needed to raise the mass 15 cm. Record this value as the effort distance in Table 2.

6. Lower the mass to the table top. As you again pull up on the spring scale, measure the distance the spring scale moves as you raise the mass a distance of 15 cm. Record this value as the effort distance in Table 2.

7. Remove the 0.5 kg mass from the pulley and the spring scale from the string.

8. Repeat steps 4–7 for the 1 kg mass and the combined 0.5 kg and 1 kg masses.

Data and Observations

1. Use Graph 1 to construct a bar graph comparing the effort force of the single fixed pulley, the effort force of the block and tackle, and the resistance force for each of the three masses. Plot the value of the masses on the x axis and the force on the y axis. Label the x axis *Mass (kg)* and the y axis *Force (N)*. Clearly label the bars that represent the values of the effort force of the single fixed pulley, the effort force of the block and tackle, and the resistance force.

2. Use Graph 2 to construct a bar graph comparing the effort distance of the single fixed pulley, the effort distance of the block and tackle, and the resistance distance for each of the three masses. Plot the value of the masses on the x axis and the distance on the y axis. Label the x axis *Mass (kg)* and the y axis *Distance (cm)*. Clearly label the bars that represent the values of the effort distance of the single fixed pulley, the effort distance of the block and tackle, and the resistance distance.

3. Work input is the work done by you. Work input can be calculated using the following equation.

$$Work\ input = Effort\ force \times Effort\ distance$$

If the force is measured in newtons (N) and the distance is measured in meters (m), work will be expressed in joules (J). Calculate the work input for the pulley and the block and tackle for each mass. Record the values in Table 3.

4. Work output is the work done by the machine. Work output can be calculated using the following equation.

$$Work\ output = Resistance\ force \times Resistance\ distance$$

If the force is measured in newtons (N) and the distance is measured in meters (m), work will be expressed in joules (J). Calculate the work output for the pulley and the block and tackle for each mass. Record the values in Table 3.

5. The efficiency of a machine is a measure of how the work output of a machine compares with the work input. The efficiency of a machine can be calculated using the following equation.

$$Efficiency = Work\ output/Work\ input \times 100\%$$

Use this equation to calculate the efficiency of the single fixed pulley and the efficiency of the block and tackle in raising each mass. Record these values in Table 4.

Laboratory Activity 1 (continued)

Table 1

Mass (kg)	Resistance force (N)	Effort force (N)	Resistance distance (cm)	Effort distance (cm)
0.5	5N	10	10 15.0	50
1.0			15.0	
1.5			15.0	

Table 2

Mass (kg)	Resistance force (N)	Effort force (N)	Resistance distance (cm)	Effort distance (cm)
0.5	5N	2 1/2	10 15.0	34
1.0			15.0	
1.5			15.0	

Table 3

Mass (kg)	Single fixed pulley		Block and tackle	
	Work input (J)	Work output (J)	Work input (J)	Work output (J)
0.5	5J	5J	85J	.5
1.0				
1.5				

Table 4

Mass (kg)	Efficiency (%)	
	Single fixed pulley	Block and tackle
0.5		
1.0		
1.5		

Laboratory Activity 1 (continued)

Graph 1

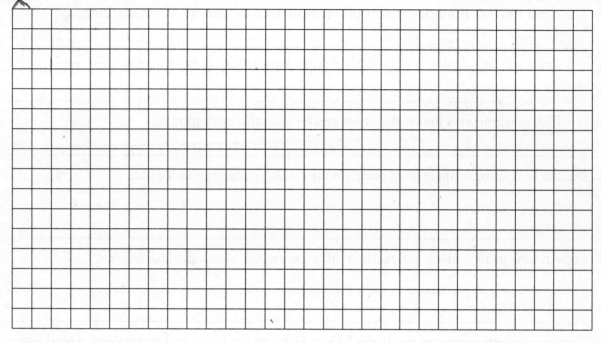

Graph 2

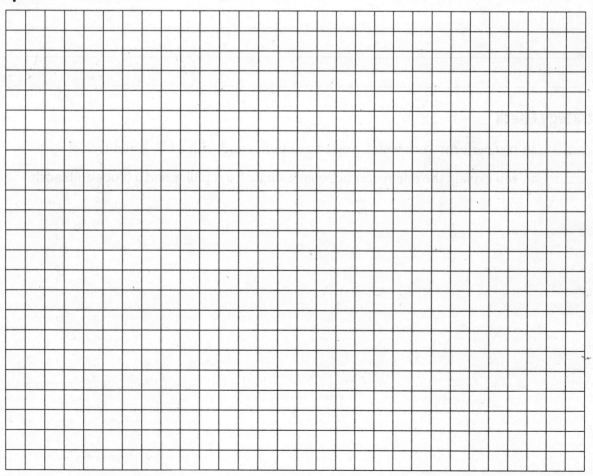

Laboratory Activity 1 (continued)

Questions and Conclusions

1. The effort distance is much greater than the resistance distance in which machines(s)?

_____

2. The effort force is much less than the resistance force in which machine(s)?

_____

3. In which machine(s) is the work output greater than the work input?

_____

4. Explain how using a single fixed pulley to raise a flag makes the task easier.

_____

_____

5. Explain how using a block and tackle to lift a car engine makes the task easier.

_____

_____

6. Compare the efficiencies of the single fixed pulley and the block and tackle. Why would you expect the block and tackle to be less efficient than the single fixed pulley?

_____

_____

_____

Strategy Check

_____ Can you perform work with a single fixed pulley and with a block and tackle?

_____ Can you explain the differences between a single fixed pulley and a block and tackle?

Causing Friction

When you kick a soccer ball along the ground, you know that when you stop kicking the ball will eventually roll to a stop. What happens to the kinetic energy of the ball as it slows down? The law of conservation of energy states that energy cannot be created or destroyed. Therefore, the kinetic energy of the soccer ball does not just disappear; it changes form. As the ball rolls over the ground, friction causes some of its kinetic energy to change into thermal energy. Friction between the ball and the ground causes the ball to slow down and eventually stop. In this experiment, you will examine how different types of surfaces affect the amount of friction produced.

Strategy

You will predict what types of surfaces produce the least friction.
You will observe how friction affects the kinetic energy of a toy car.

Materials

books (2)   coarse sandpaper (3 sheets)
meterstick   strip of rough carpeting
toy car   pillowcase
masking tape

Procedure

1. Place a book on top of a smooth, hard surface such as a table or an uncarpeted floor. Lean a second book against the first to form a ramp.

Figure 1

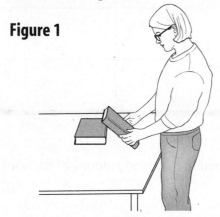

2. Use the meterstick to measure the height of the ramp. See Figure 1. Record the height in the Data and Observations section.

3. Do you think hard surfaces or soft surfaces will reduce the kinetic energy of a toy car more quickly? Rough surfaces or smooth surfaces? Record your predictions in the Data and Observations section.

4. Place the car at the top of the ramp and release it. Measure the distance between the bottom of the ramp and the spot where the car stopped moving. Record this distance in the table in the Data and Observations section. Repeat this step two more times. Use the meterstick to make sure that you release the car from the same height each time.

5. Tape the pieces of sandpaper together into a strip. Place the strip at the bottom of your ramp. See Figure 2. Repeat Step 4.

Figure 2

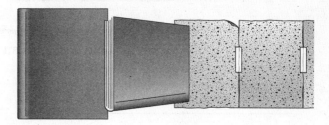

Laboratory Activity 2 (continued)

6. Use your fingers to brush ridges into the carpet's surface. Remove the sandpaper from the bottom of your ramp and replace it with the carpeting. Repeat Step 4.

7. Fold the pillowcase lengthwise into thirds. Place it on top of the carpeting at the bottom of your ramp.
   Smooth out any wrinkles in the fabric with your hands. Repeat Step 4.

8. Calculate the average distance the car traveled on each surface. Record your calculations in the Data Table.

Data and Observations

Height of ramp: _____

Predict what type of surfaces—hard or soft, smooth or rough—will reduce the kinetic energy of the car the quickest?

_____
_____

Surface	Distance Moved by Car (cm)			Distance Moved by Car (cm)
	Trial 1	Trial 2	Trial 3	
Floor or table (hard, smooth)				
Floor or table (hard, rough)				
Floor or table (soft, rough)				
Pillowcase (soft, smooth)				

Questions and Conclusions

1. What type of surface (hard or soft, smooth or rough) provided the greatest amount of friction? Explain how you know.

_____
_____

2. What type of surface provided the least amount of friction?

_____
_____

Laboratory Activity 2 (continued)

3. What happened to the kinetic energy of the car after the car left the ramp?

_____

_____

_____

4. Why was it important that the ramp be the same height in each trial?

_____

_____

_____

_____

5. Describe how you could determine the gravitational potential energy of the car at the top of the ramp.

_____

_____

_____

6. Examine the data in your table. Then predict the distance the car would move if you placed a layer of gravel at the bottom of your ramp. Explain how the data helped you make your prediction.

_____

_____

_____

7. Predict whether a hockey puck would move a greater distance over smooth ice or over rough ice. Explain how you used the data in your table to make your prediction.

_____

_____

_____

Strategy Check

_____ Can you predict what types of surfaces will produce the least friction?

_____ Can you observe how friction affects kinetic energy?

Specific Heats of Metals

The amount of heat needed to change the temperature of a metal is much less than that needed to change the temperature of a similar amount of other materials. You probably were aware of this fact if you ever tried to cool a can of soft drink quickly in the freezer. Metal cans tend to cool more quickly than their contents.

A measure of how much energy is needed to change the temperature of a material is called specific heat. The specific heat, C, is the amount of heat needed to change the temperature of 1 kilogram of a substance by 1 degree Celsius. As you recall, the specific heat of water is 4190 J/kg•°C. The specific heat of a substance is a physical property of that substance. Therefore, a substance can be identified by its specific heat.

Strategy

You will use a calorimeter to determine the specific heat of a piece of metal.
You will identify the metal by its specific heat.

Materials

250-mL beaker	plastic pipette	test-tube rack
one-hole paper punch	rubber band	thermometer
metric balance	test tube, thick walled	sample of unknown metal X, Y, or Z
paper towels	test-tube holder	water
2 plastic cups with lids		

Procedure

Part A—Building a Calorimeter

1. Place about 50 mL of water in the 250-mL beaker and allow the temperature of the water to come to room temperature.
2. Punch a hole for the thermometer in one of the lids of the plastic cups with a paper punch.
3. Wrap a rubber band around one of the plastic cups.

4. Place this cup inside the second plastic cup. Assemble the calorimeter as shown in Figure 1.
5. Measure the mass of the calorimeter. Record this value in the Data and Observations section.

Figure 1

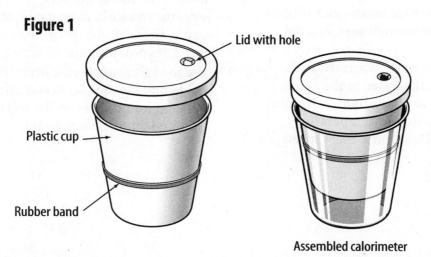

Lid with hole

Plastic cup

Rubber band

Assembled calorimeter

Laboratory Activity 1 (continued)

Part B—Measuring Specific Heat of Metals

1. Use the plastic pipette to add 5 pipette-fuls of room-temperature water to the calorimeter.

2. Measure the mass of the calorimeter and water. Record this value in the Data and Observations section.

3. Measure the mass of the sample of unknown metal. Record this value in Table 1.

4. Place the piece of metal in the test tube. Use the test-tube holder to place the test tube containing the metal into the boiling water bath prepared by your teacher. Note the time.

5. After ten minutes, measure the temperature of the water in the calorimeter with the thermometer. Remove the thermometer. Place the calorimeter on a paper towel on a flat surface and remove its lid.

6. Measure the temperature of the boiling water bath using the thermometer provided by your teacher. Record this value as the *temperature of metal* in the Data and Observations section.

7. Using the test-tube holder, carefully remove the test tube containing the sample from the boiling water bath. **CAUTION:** *The test tube and its contents are extremely hot. Avoid touching the test tube or the piece of metal.*

8. Quickly slide the piece of hot metal into the calorimeter. Place the test tube in the test-tube rack.

9. Immediately cover the calorimeter with its lid and insert the thermometer into the calorimeter.

10. Gently swirl the water. Measure the temperature of the water in the calorimeter for several minutes. Record the value of the *highest* temperature reading in the Data and Observations section.

11. Calculate the mass of the water that you added to the calorimeter by subtracting the mass of the empty calorimeter from the calorimeter and water. Record this value in Table 1.

12. Calculate the temperature change in the water. Record this value in Table 1.

13. The heat gained by the water can be determined by the following equation.
$$Q = C \times m \times (T_f - T_i)$$
In this equation, C represents the specific heat of water, m represents the mass of the water, and T_f is the final temperature and T_i is the initial temperature of the water. Calculate the value of Q and record it in Table 1.

14. Assume that all the heat from the metal was transferred to the water in the calorimeter. Thus, the heat lost by the metal is equal to the heat gained by the water. Enter the value of the heat lost by the metal in Table 1. Remember to record a heat loss as a negative value.

15. Calculate the change in temperature of the metal.

16. The specific heat of a substance can be calculated by the following equation.
$$C = \frac{Q}{m(T_f - T_i)}$$
In this equation, Q represents the amount of heat gained or lost, m represents the mass of the substance, and $(T_f - T_i)$ represents the change in temperature of the substance. Calculate the specific heat of the metal. Record this value in Table 1.

17. Use the values of specific heats in Table 2 to identify the sample. Record the letter of the sample and name of the metal in the Data and Observations section.

Laboratory Activity 1 (continued)

Data and Observations

Mass of calorimeter: _____ g

Mass of calorimeter and water: _____ g

Temperature of cool water: _____ °C

Temperature of metal: _____ °C

Temperature of water-metal mixture: _____ °C

Table 1

Metal	Specific heat (J/kg·C°)

Table 2

Measurement/Calculation	Material	
	Water	Metal
Mass (kg)		
Temperature change (°C)		
Specific heat (J/kg · °C)		
Heat gained or heat lost (J)		

Sample _____; Name of metal: _____

Questions and Conclusions

1. How well were you able to identify the metal using its specific heat?

_____

_____

2. In this experiment, the masses of the metal and the hot water were almost equal. However, the temperature decrease of the metal was much greater than the temperature rise of the water even though they had equal masses. Why?

_____

_____

_____

Laboratory Activity 1 (continued)

3. In step 11 of the procedure, you recorded the temperature of the water bath as the temperature of the metal in it. Explain why you could do this.

_____

_____

_____

_____

4. Could you improve your calorimeter by using two metal cans and aluminum foil in place of the two plastic cups and lids? Explain.

_____

_____

_____

_____

Strategy Check

_____ Can you find the specific heat of a metal?

_____ Can you identify a metal when given its specific heat?

Thermal Energy from Foods

Chapter 5

LAB 2 Laboratory Activity

You use food as fuel for your body. Food contains the stored energy you need to be active, both mentally and physically. To keep your body processes going, your body must release the energy stored in food by digesting the food.

You cannot directly measure the energy contained in food. However, you can determine the amount of thermal energy released as a sample of food is burned by determining the thermal energy absorbed by water heated by the burning sample. By measuring the temperature change of a given mass of water, you can calculate the energy released from the food sample. Raising the temperature of 1 kg of water by 1 Celsius degree requires 4,190 joules of energy. This information can be expressed as the specific heat (C) of water, which is 4,190J/kg • °C. You can use the following equation to determine the heat (Q) released when a food sample is burned.

$$energy\ released = energy\ absorbed$$
$$energy\ absorbed = temperature\ change\ of$$
$$water \times mass\ of\ water \times specific\ heat\ of\ water$$
$$Q = (T_f - T_i) \times m \times C$$

Strategy

You will calculate a change in thermal energy.

You will account for the difference between energy released and energy absorbed.

Materials

large paper clip or long pin
food sample
aluminum potpie pan
metric balance
100-mL graduated cylinder

water
100-mL flask
utility clamp
ring stand

thermometer
wood splint
matches
watch or clock

Procedure

1. Wear a laboratory apron and safety goggles throughout this experiment. Straighten the paper clip and insert it through the food sample. Position the paper clip on the edges of the aluminum potpie pan as shown in Figure 1. Use the balance to determine the mass of the pan, paper clip, and food sample. Record the mass in Table 1 as m_1.

2. Use the graduated cylinder to add 50 mL of water to the flask. Clamp the flask on the ring stand about 5 cm above the tabletop. Use the thermometer to measure the temperature of the water. Record this value in Table 1 as T_i.

3. Ignite the wood splint with a match. **CAUTION:** *Always use care with fire.* Use the burning splint to ignite the food sample. Once the food sample is burning, safely extinguish the splint. Position the aluminum pan under the flask. The water in the flask should absorb most of the energy released by the burning food.

Figure 1

Laboratory Activity 2 (continued)

4. Stir the water with the thermometer and closely observe the temperature rise.
5. Blow out the flame of the burning food after about 2 minutes. Record the highest temperature of the water during the 2 minutes in Table 1 as T_f.
6. Allow the aluminum pan and its contents to cool. Determine the mass of the pan and contents after the release of energy. Record this value in Table 1 as m_2.

Figure 2

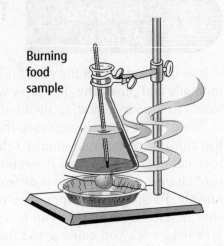

Burning food sample

Data and Observations

Table 1

Food sample	Mass (kg)		Temperature (°C)	
	m_1	m_2	T_i	T_f

$T_f - T_i =$ _____

m (mass of 50 mL of water) = _____

$(m_2 - m_1) =$ _____

$Q =$ _____

Heat absorbed per gram of food burned = _____

1. Calculate the rise in the water temperature by subtracting T_i from T_f. Record this value.
2. Use the equation given in the introduction to calculate the energy absorbed by the water when the food sample was burned. Be sure to use the mass of the water for m. Record this value.
3. Calculate the heat absorbed per gram of food by dividing the energy absorbed by the water by the mass of food burned $(m_2 - m_1)$. Record this value.
4. Your teacher will make a data table of food samples and energy absorbed by the water in the flask. Record your data in this table.

Laboratory Activity 2 (continued)

Questions and Conclusions

1. In order to calculate the amount of energy released or absorbed by a substance, what information do you need?

_____

_____

2. How do you know that energy was transferred in this experiment?

_____

3. Did you measure the energy released by the food sample or the energy gained by the water?

_____

_____

4. Most of the energy of the burning food was absorbed by the water. What do you think happened to the small amount of energy that was not absorbed by the water?

_____

_____

5. Look at the data table of different food samples tested by your class. Which food sample released the most energy? Which food sample released the least energy?

_____

_____

6. Suppose 20.0g of your food sample is burned completely. Use a proportion to calculate the value of energy released.

_____

_____

Strategy Check

_____ Can you calculate a change in thermal energy?

_____ Can you determine whether energy is released or absorbed?

Wet Cell Battery

A car battery consists of a series of wet cells. Each wet cell contains two plates called electrodes, made of different metals or metallic compounds, and a solution called an electrolyte. Chemical reactions occur between the electrodes and the electrolyte. These reactions create a voltage difference between the two electrodes. Voltage difference is measured in a unit called the volt (V). If the two electrodes of a wet cell are connected by a conductor, electrons will flow through the conductor from one electrode, called the negative (−) electrode, to the other, called the positive (+) electrode. Within the cell, electrons will flow from the positive electrode to the negative electrode. The flow of electrons is caused by a chemical reaction.

Wet cells vary in their voltage difference. The voltage difference of a wet cell depends on the materials that make up the electrodes.

Strategy

You will construct a wet cell.

You will measure the voltage difference of a wet cell with a voltmeter.

You will observe how the voltage difference of a cell depends on the electrode materials.

Materials

2 alligator clips
250-mL beaker
long iron nail
100-mL graduated cylinder

paper towels
2 wires
tin strip
zinc strip

short wire tie
hydrochloric acid (HCl)
2 glass rods
voltmeter

Procedure

1. Place two glass rods across the top of the beaker.
2. Use an alligator clip to hang the zinc strip from one of the glass rods. The strip should hang near one side of the beaker.
3. Attach one wire to the alligator clip.

Attach the other end of the wire to the negative (−) terminal of the voltmeter.

4. Attach the iron nail to the second glass rod with the small wire tie. Attach the second alligator clip to the top of the nail. See Figure 1.
5. Connect the second alligator clip to the positive (+) terminal of the voltmeter with the other wire as shown in Figure 2.

Figure 1

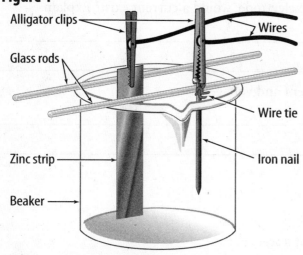

Figure 2

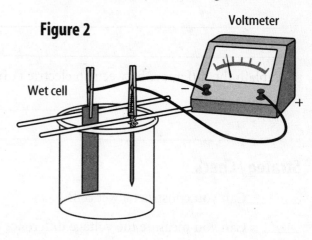

Laboratory Activity 1 (continued)

6. Carefully add 75 mL of hydrochloric acid to the beaker. **CAUTION:** *Hydrochloric acid causes burns. Rinse any spills immediately with water.* Make sure that the zinc strip and the nail are partially submerged in the acid.

7. Observe the wet cell. Record any changes in **Table 1**. Record the reading of the voltmeter in the data table.

8. Disconnect the wires. Carefully empty the acid from the beaker where your teacher indicates. Rinse the beaker, zinc strip, and iron nail and dry them with paper towels.

9. Repeat steps 1 through 8 using the zinc strip and the tin strip. In step 4, attach the tin strip to the glass rod with the alligator clip. After adding the HCl to the cell, record your observations and the reading of the voltmeter in Table 1.

Data and Observations

Table 1

Electrodes	Observations	Voltage difference (V)
zinc, iron		
zinc, tin		

Questions and Conclusions

1. How do you know that a chemical reaction has occurred in the wet cell after you added the acid?

_____

2. Which pair of electrodes produced the greater voltage difference?

_____

3. If one of the alligator clips is removed from the electrode, would a current exist? Explain.

_____

_____

4. Explain the difference between an electric current and voltage.

_____

_____

Strategy Check

_____ Can you construct a wet cell?

_____ Can you measure the voltage difference of a wet cell?

Simple Circuits

LAB **2** Laboratory Activity

Can you imagine a world without electricity? It is hard to believe that electrical energy became commercially available in the early 1880s.

The appliances plugged into the wall outlets of a room are part of an electric circuit. The most simple type of electric circuit contains three elements:

- a source of electrical energy, such as a dry cell;
- a conductor such as copper wire, which conducts an electric current; and
- a device, such as a lamp, which converts electrical energy into other forms of energy.

Complex circuits may contain many elements. How the elements are arranged in a circuit determines the amount of current in each part of a circuit.

Strategy

You will construct a series circuit and a parallel circuit.
You will observe the characteristics of the elements in circuits.
You will compare and contrast the characteristics of elements in series and parallel circuits.

Materials

aluminum foil
20-cm × 20-cm cardboard sheet
2 LEDs (light-emitting diodes)
metric ruler

9-V dry cell battery
9-V mini-battery clip
500-Ω resistor

scissors
stapler and staples
transparent tape

Procedure

Part A—Constructing and Observing a Series Circuit

1. Place the cardboard sheet on a flat surface.
2. Cut 2 1-cm × 10-cm strips of aluminum foil with the scissors.
3. Attach the battery clip to the 9-V mini-battery. Securely attach the battery and the two aluminum foil strips to the board with tape as shown in Figure 1.

4. Staple the exposed end of the red lead wire from the battery clip to the top foil strip. Staple the exposed end of the black lead wire from the clip to the bottom foil strip as shown in Figure 2. Be sure that the staples are pressing the exposed ends of the wires securely against the foil strips.

Figure 1

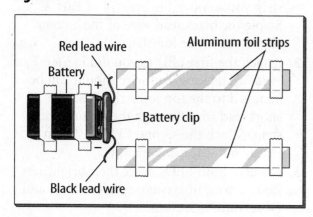

Figure 2

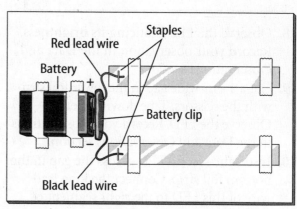

Laboratory Activity 2 (continued)

5. Cut a 1.0-cm-wide gap in the top foil strip with the scissors. Tape down the ends as shown in Figure 3.

6. Place the 500-Ω resistor across the gap. Securely staple the two wires of the resistor to the cut aluminum strip as shown in Figure 3.

Figure 3

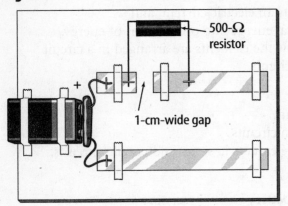

7. Push the long lead wire of the LED into the top aluminum strip. Push the short lead wire from the LED into the bottom strip as shown in Figure 4.

Figure 4

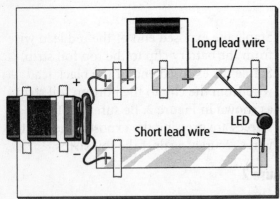

8. Observe the LED, noticing its brightness. Record your observation in the Data and Observations section.

9. Cut a 1-cm-wide gap in the lower foil strip with the scissors. Tape down the ends. Observe the LED. Record your observations in the Data and Observations section.

10. Insert the second LED across the gap in the bottom foil strip. Connect the long lead wire of this LED to the right segment of the strip as shown in Figure 5.

Figure 5

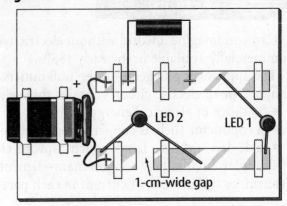

Attach the short lead wire to the left segment of the foil strip.

11. Observe both LEDs. Note if the brightness of LED 1 has changed from step 8. Record your observations in the Data and Observations section.

12. Predict what will happen to LED 2 if LED 1 is removed. Record your prediction.

13. Remove the first LED and observe the second LED. Record your observations.

14. Carefully remove LED 2, the staple from the black lead wire of the battery clip, and the two segments of the bottom foil strip from the cardboard sheet. (Disconnect LED 1 from the bottom foil strip first.) Leave all other circuit elements attached to the cardboard sheet for Part B of the experiment.

Part B—Constructing and Observing a Parallel Circuit

1. Cut a 1-cm × 10-cm strip of aluminum foil. Tape it to the board in place of the strip you removed in Step 14 of Part A. Staple the black lead wire of the battery clip to the lower foil strip.

2. Attach the first LED as you did in Step 7 of Part A. The long lead wire should still be attached to the top foil strip. Push the short lead wire through the bottom foil strip. Attach the second LED as shown in Figure 6 in the same manner.

3. Observe both LEDs. Note their brightness. Record your observations in the Data and Observations section.

Laboratory Activity 2 (continued)

4. Predict what will happen if LED 1 is removed. Record your prediction.
5. Remove LED 1. Record your observations.
6. Replace LED 2 and observe both LEDs. Note any change in brightness of the LEDs. Record your observations.
7. Predict what will happen if LED 2 is removed. Record your prediction.
8. Remove LED 2 and observe LED 1. Record your observations.

Figure 6

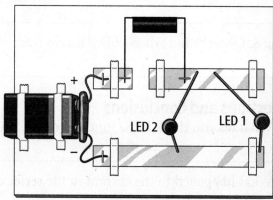

Data and Observations

Because the brightness of an LED in a circuit is directly related to the current in the circuit, the brightness of the LED is a measure of the current in that part of the circuit containing the LED.

Part A—Constructing and Observing a Series Circuit

Step 8. Observation of the LED when inserted into the foil strips:

_____

Step 9. Observation of the LED when lower foil strip is cut:

_____

Step 10. Observation of LEDs 1and 2 when LED 2 is inserted across gap in bottom foil strip:

_____

Step 11. Prediction if LED 1 is removed:

_____

Step 12. Observation when LED 1 is removed:

_____

Part B—Constructing and Observing a Series Circuit

Step 3. Observation of LEDs 1 and 2:

_____

Step 4. Prediction if LED 1 is removed:

_____

Step 5. Observation when LED 1is removed:

_____

Step 6. Observation when LED 2 is replaced:

_____

Laboratory Activity 2 (continued)

Step 7. Prediction if LED 2 is removed:

_____

Step 8. Observation when LED 2 is removed:

_____

Questions and Conclusions

1. What do you think is the function of the 500-Ω resistor?

_____

2. What happened to the current in the series circuit when an LED was removed?

_____

3. What happened to the current in the series circuit when another LED was added?

_____

4. What happened to the current in the parallel circuit when an LED was removed?

_____

5. What happened to the current in the first LED in your parallel circuit when the second LED was added?

_____

6. Explain what your answer to question 4 indicates about the total amount of current in the resistor.

_____

7. How do you know if the lamps plugged into wall outlets in your house are part of a series circuit or a parallel circuit?

_____

_____

_____

Strategy Check

_____ Can you construct a series circuit?

_____ Did your observations reflect your predictions?

Comparing Magnetic Fields

A magnetic material is made of small regions called magnetic domains. These magnetic domains can be pictured as small bar magnets. When the domains are aligned, as shown in Figure 1, the magnetic fields of the domain add together. This causes the material to be surrounded by a magnetic field.

The magnetic field surrounding a magnet exerts a magnetic force on other magnets and magnetic materials. The direction of the magnetic field around a magnet can be represented by magnetic field lines. Magnetic field lines always begin on the north pole of a magnet and end on the south pole. Magnetic field lines are closer together where the magnetic field is stronger, and farther apart where the field is weaker.

Figure 1

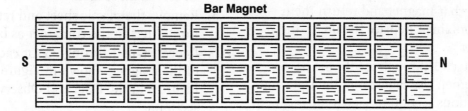

Bar Magnet

S   N

Strategy

You will observe the effect of a magnetic field around a magnet.
You will represent the shape of magnetic field lines by drawing an example.
You will compare and contrast the magnetic field lines around a bar magnet and a horseshoe magnet.
You will observe the interaction of two magnetic fields.

Materials

sheet of clear plastic
cardboard frame
masking tape
short bar magnets (2)
iron filings in a plastic container with a shaker top
small horseshoe magnet

Procedure

Part A—Magnetic Field of a Magnet

1. Attach the plastic sheet to the cardboard frame with masking tape.
2. Lay one bar magnet on a flat surface with its north pole at the left. Place the frame over the magnet so that the magnet is centered within the frame as shown in Figure 2.

Figure 2

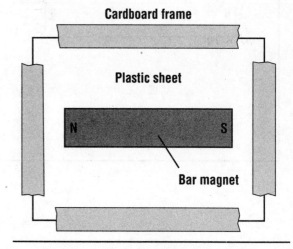

Cardboard frame

Plastic sheet

N   S

Bar magnet

Laboratory Activity 1 (continued)

3. Gently sprinkle iron filings onto the plastic sheet. Observe how the magnetic field of the magnet affects the iron filings. The iron filings line up along the magnetic field lines around the bar magnet.

4. Sketch the magnetic field lines around the bar magnet in Figure 3 in the Data and Observations section.

5. Remove the lid from the container of the iron filings. Remove the tape holding the plastic sheet to the frame. Carefully lift the sheet and pour the iron filings into the container. Pick up any spilled filings with the other bar magnet and return them to the container. Replace the lid on the container.

6. Repeat steps 1 through 5 with the horseshoe magnet. Use Figure 4 in the Data and Observations section to sketch the magnetic field lines around the horseshoe magnet.

Part B—Interaction of Magnetic Fields

1. Attach the plastic sheet to the cardboard frame with masking tape.

2. Lay two bar magnets end to end on a flat surface as shown in Figure 5 in the Data and Observations section. Place the frame over the magnets so that they are centered within the frame.

3. Gently sprinkle iron filings onto the plastic sheet.

4. Sketch the magnetic field lines around the two bar magnets in Figure 5 in the Data and Observations section.

5. Remove the plastic sheet and return the iron filings to the container as before.

6. Repeat steps 1 through 5 for each position of the magnets shown in Figure 6 through Figure 8 in the Data and Observations section.

Data and Observations

Part A—Magnetic Field of a Magnet

Figure 3

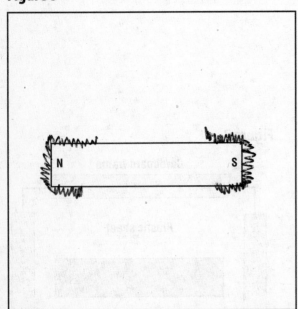

Figure 4

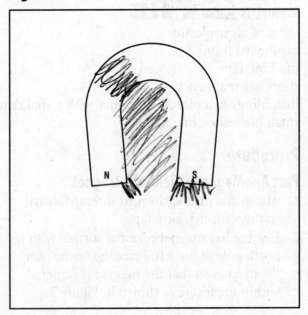

Laboratory Activity 1 (continued)

Part B—Interaction of Magnetic Fields

Figure 5

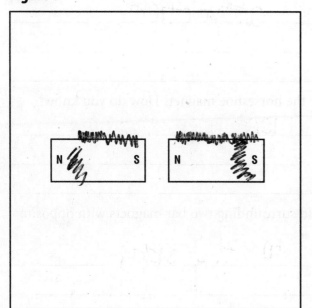

Figure 6

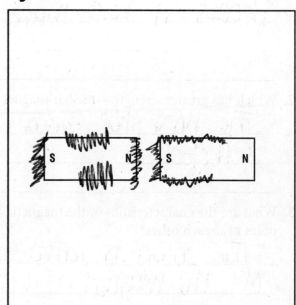

Figure 7

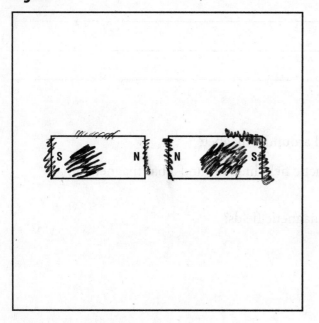

Figure 8

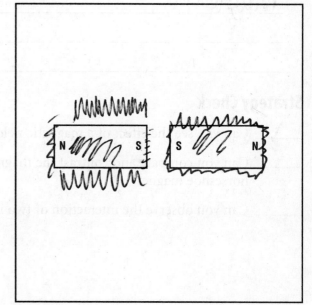

Laboratory Activity 1 (continued)

Questions and Conclusions

1. Why were you able to see the magnetic field lines using iron filings?

Because it had magnetic attraction

2. Which has greater strength—the bar magnet or the horseshoe magnet? How do you know?

The horse shoe because more iron was collected

3. What are the characteristics of the magnetic field surrounding two bar magnets with opposite poles near each other?

The iron is attracted to the sides of the magnet.

4. What are the characteristics of the magnetic field surrounding two bar magnets with like poles near each other?

Theres more iron in the center of the magnetic

Strategy Check

____✓____ Can you see the effect of a magnetic field around a magnet?

____✓____ Can you compare and contrast the magnetic field lines around a bar magnet and a horseshoe magnet?

____✓____ Can you observe the interaction of two magnetic fields?

Creating Electromagnets

A magnetic field exists around any wire that carries an electric current. By coiling the wire around a bolt or nail, the strength of the magnetic field will increase. A coil of wire wrapped around a bolt or nail will become an electromagnet if the wire is connected to a battery or other source of current. The magnetic force exerted by an electromagnet can be controlled by changing the electric current.

Strategy

You will construct several electromagnets.

You will compare the strength of the magnetic force of four electromagnets.

You will determine the relationship between the strength of the magnetic force and the number of turns of wire in the coil of the electromagnet.

Materials

iron bolts, identical, at least 5 cm long (4)
marking pen
masking tape
BBs, iron
*paper clips
small, plastic cups (2)
insulated wire
1.5-V dry cell battery
*Alternate materials

Procedure

1. Place masking tape on the heads of the bolts and label the bolts *A, B, C,* and *D*.
2. Put all the BBs in one cup.
3. Test each bolt for magnetic properties by attempting to pick up some of the BBs from the cup. Record your observations in the Data and Observations section.
4. Wrap 10 full turns of wire around bolt A. Wrap 20 turns of wire around bolt B, 30 turns around bolt C, and 40 turns around bolt D.

5. Use masking tape to connect the ends of the wires of bolt A to the dry cell as shown in the figure. Carefully use your electromagnet to pick up as many BBs as possible. Hold the electromagnet with the BBs over the empty cup and disconnect the wire to the dry cell. Make sure all the BBs fall into the cup. Count the number of BBs in the cup. Record this value in the table in the Data and Observations section.

Figure 1

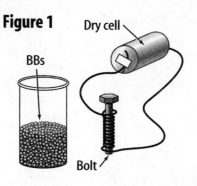

Dry cell

BBs

Bolt

Copyright © Glencoe/McGraw-Hill, a division of The McGraw-Hill Companies, Inc.

Laboratory Activity 2 (continued)

6. Return all the BBs to the first cup.
7. Repeat steps 5 and 6 using bolts B, C, and D. Record in the table the number of BBs each electromagnet picked up.
8. Use the blank graph in the Data and Observations section to construct a graph relating the number of BBs picked up by the electromagnet and the number of turns of wire in the electromagnet.

Determine which axis should be labeled *Number of BBs picked up* and which should be labeled *Number of turns of wire*. Plot your findings on the graph.

Data and Observations

Observation of the magnetic properties of the bolts alone:

_____

_____

_____

Electromagnet	Number of turns of wire	Number of BBs picked up
A	10	
B	20	
C	30	
D	40	

Laboratory Activity 2 (continued)

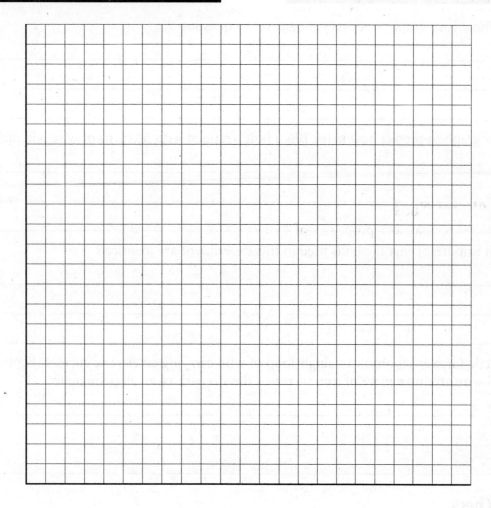

Questions and Conclusions

1. How is the number of BBs that were picked up related to the magnetic force?

_____

_____

_____

2. How is the strength of the magnetic force exerted by an electromagnet related to the number of turns of wire?

_____

_____

_____

_____

Laboratory Activity 2 (continued)

3. Explain how your graph supports your answer to question 2.

_____

_____

_____

4. Use your graph to predict how many BBs a bolt wrapped with 50 turns of wire will pick up.

_____

_____

_____

5. Why is it important that the bolts used in this experiment are identical?

_____

_____

_____

6. A magnetic force exists around a single loop of wire carrying an electric current. Explain why coiling a wire around a piece of iron increases the strength of an electromagnet.

_____

_____

_____

Strategy Check

_____ Can you construct electromagnets of different strengths?

_____ Can you compare the strength of the magnetic force exerted by different electromagnets?

_____ Can you explain how the strength of the magnetic force is related to the number of turns of wire in the coil of an electromagnet?

Solar Cells

The Sun's radiant energy drives the weather and water cycles of Earth. This energy is necessary to sustain life on Earth. It might also be powering your pocket calculator or providing the hot water for your next shower.

Many pocket calculators contain solar cells. A solar cell is a device that converts radiant energy into electrical energy. In a circuit, a solar cell can produce an electric current. In this experiment you will investigate the power output of solar cells.

Strategy

You will determine the power of output of a solar cell.

You will describe how the power output of a solar cell is related to the power rating of its energy source.

You will compare sunlight and artificial sources of radiant energy.

Materials 🔌 🥽

25-, 60-, 75-, and 100-W lightbulbs
10-cm lengths of insulated wire (4)
light socket and cord
utility clamp
ring stand
meterstick
masking tape
solar cell
DC voltmeter
DC ammeter
resistor switch

Procedure

Part A—Artificial Sources of Light

1. Place the 25-W lightbulb into the light socket.

2. Attach the utility clamp to the ring stand. Use the utility clamp to position the light socket so that the bulb is 50 cm above the desk top. **CAUTION:** *Tape the socket's electrical cord onto the desk top so that no one can trip over the cord or topple the ring stand.*

3. Place the solar cell parallel to the desk and directly beneath the bulb.

4. Connect the voltmeter, ammeter, switch, and solar cell with the insulated wires as shown in Figure 1.

5. Plug the socket cord into an electrical outlet. Darken the room.

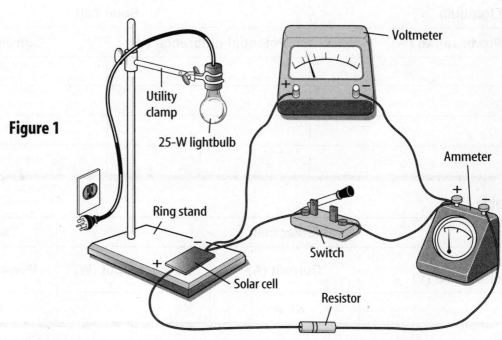

Figure 1

Laboratory Activity 1 (continued)

6. Switch on the 25-W bulb. Close the switch in the circuit. Measure the potential difference and the current with the voltmeter and ammeter, respectively. Record the value of the power rating of the bulb, the potential difference, and current in Table 1.

7. Open the circuit switch. Turn off the bulb and allow it to cool. **CAUTION:** *Lightbulbs generate heat. Do not touch the lightbulb for several minutes.*

8. Remove the lightbulb and replace it with the 60-W bulb.

9. Repeat steps 6–8 for the 60-W, 75-W, and 100-W lightbulbs.

Part B—Sunlight

1. Move the circuit containing the solar cell, voltmeter, ammeter, and switch to a sunny location, such as a window sill. Position the solar cell perpendicular to the sunlight.

2. Close the circuit switch. Measure the potential difference and the current with the voltmeter and ammeter, respectively. Record these values in Table 2. Open the switch.

Analysis

1. The power output (P) of the solar cell can be calculated using the following equation.
$$P = V \times I$$
In this equation V represents the potential difference measured in volts (V) and I represents the current measured in amperes (A). The unit of power is the watt (W). Use this equation to calculate the power output of the solar cell for each lightbulb. Record in Table 3 the value of the power rating of the lightbulb and the power output of the solar cell for each lightbulb.

2. Use Graph 1 to plot the power rating of the lightbulbs and the power output of the solar cell. Determine which variable should be represented by each axis.

3. Calculate the power output of the solar cell from Part B of the procedure. Record the value in Table 2.

4. Use Graph 1 to estimate the power rating of sunlight. Record the value in Table 2.

Data and Observations

Table 1

Lightbulb	Solar cell	
Power rating (W)	Potential difference (V)	Current (A)

Table 2

Solar cell			Sunlight
Potential difference (V)	Current (A)	Power output (W)	Power rating (W)

Laboratory Activity 1 (continued)

Table 3

Lightbulb	Solar cell
Power rating (W)	Power output (W)

Graph 1

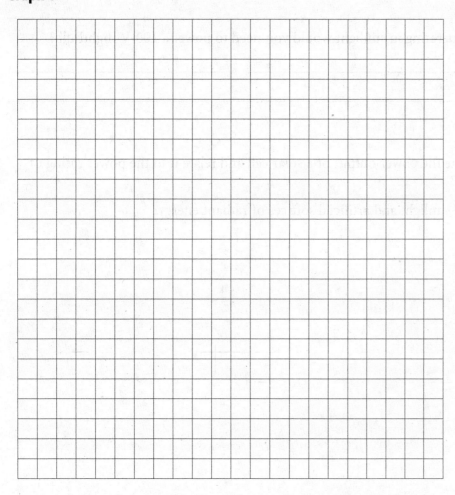

Laboratory Activity 1 (continued)

Questions and Conclusions

1. Which lightbulb produced the greatest power output of the solar cell?

_____

2. How is the power output of the solar cell related to the power rating of the lightbulbs?

_____

_____

_____

3. How does sunlight compare to artificial sources of light?

_____

_____

4. How many solar cells operating in sunlight would you need to power a 100-W lightbulb?

_____

_____

Strategy Check

_____ Can you determine the power output of a solar cell and relate it to the power rating of the energy source?

_____ Can you compare sunlight and artificial sources of radiant energy?

LAB 2 Laboratory Activity

Using the Sun's Energy

You may recall how water in a garden hose lying in the grass can become hot on a sunny afternoon. Allowing the Sun's radiant energy to warm water in a solar collector is one way people are using solar energy to heat homes. To be useful and efficient, the solar collector must absorb and store a large amount of solar energy. In this experiment you will see how a solar collector can be used to heat water.

Strategy

You will build a solar hot water heater.

You will measure the temperature change of the heated water.

You will explain some benefits and problems in using solar heat.

Materials

100-mL graduated cylinder	black cloth or paper
water	tape
plastic foam cup	black rubber or plastic tubing, 5–6 m
pen or pencil	buckets (2)
scissors	clothespin, spring-loaded
shallow box	thermometer
metric ruler	graph paper

Procedure

Part A—Building a Solar Water Heater

1. Use the graduated cylinder to add 100 mL of water to the plastic foam cup. Use the pencil or pen to mark the surface of the water on the inside of the cup. (Do not use a felt tip marker.) Discard the water and save the cup for later use.

2. Make 2 holes near the bottom of a large shallow box as shown in Figure 1. The diameter of each hole should be the same as the diameter of the outside of the rubber tubing. Label one hole *IN* and the other *OUT*.

3. Line the box with a black cloth or paper. If paper is used, tape it securely in place. The top of the box must be open to the sun.

4. Fold the rubber tubing in place inside the box as shown. Arrange the tubing so most of it will be exposed to the sun. The ends of the tubing should extend from the holes in the box.

Figure 1

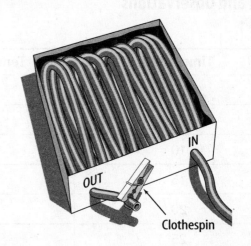

OUT

IN

Clothespin

Laboratory Activity 2 (continued)

Part B—Using Solar Energy

1. Move the box to a sunny location. Turn the box so that it is in direct sunlight.

2. Place an empty bucket beneath the tubing leading from the *OUT* hole. Place a second bucket filled to the top with water so it is above the level of the box. See Figure 2. Shade the bucket of water from the sun. Your teacher will show you how to start a siphon to fill the tube.

3. When the entire tube is filled with water, pinch the *OUT* tube with a spring-loaded clothespin. The flow of water should stop. Maintain the siphon. Do not remove the *IN* tube from the bucket of water.

4. Slowly release the clothespin and fill the plastic foam cup to the 100 mL line. New water should siphon into the system through the *IN* tube. Measure the temperature of the water in the cup with the thermometer. Record the temperature in Table 1 as the temperature at 0 minutes.

5. Collect samples of water from the water heater every 5 minutes. Check to make sure new water is siphoning into the system from the bucket. Measure and record the temperature of each water sample in Table 1.

Figure 2

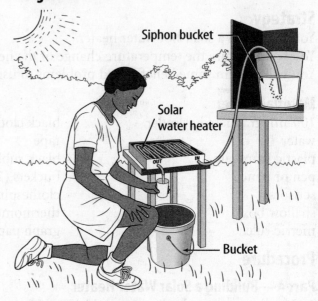

Siphon bucket

Solar
water heater

Bucket

Data and Observations

Table 1

Time (min)	Temperature (°C)
0	
5	
10	
15	
20	
25	
30	

Laboratory Activity 2 (continued)

Use Graph 1 to graph the time and temperature of the water that you heated in the solar water heater. Determine which variable should be represented by each axis.

Graph 1

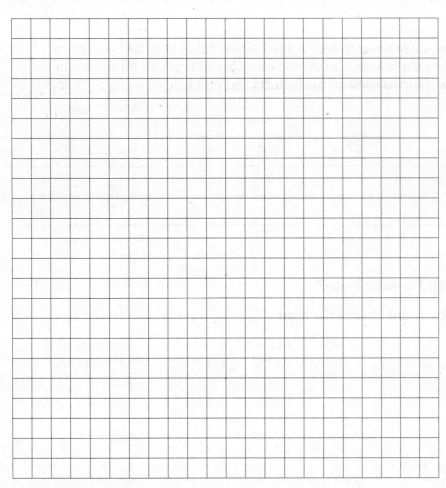

Questions and Conclusions

1. What happens to the temperature of the water in the tubing as it is exposed to the Sun?

_____

_____

_____

2. Explain how your graph indicates that solar energy can be used to heat water.

_____

_____

_____

Laboratory Activity 2 (continued)

3. Why is the inside of the box of the solar water heater black?

_____

_____

_____

4. Designers are using solar-heated water to heat entire houses. Tubes of heated water run through the walls of these solar houses. Usually the water heater is placed on the top of the house in a sunny location. Discuss some of the benefits and problems of using solar energy in this way to heat a house.

_____

_____

_____

_____

Strategy Check

_____ Can you build a solar hot water heater?

_____ Can you evaluate the usefulness of solar heat?

Velocity of a Wave

Energy can move as waves through material such as ropes, springs, air, and water. Waves that need a material to pass through are called mechanical waves. Ripples in flags and sound waves are examples of mechanical waves. Electromagnetic waves, such as light, can be transmitted through matter as well as empty spaces.

The high part or hill of a transverse wave is the crest. The low part or valley of a transverse wave is the trough. The amplitude of a mechanical wave is the distance the material through which the wave is passing rises or falls below its usual rest position. Mechanical waves of large amplitude transmit more energy than mechanical waves of small amplitude.

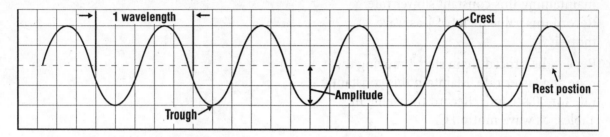

The wavelength is the distance between two similar points on successive waves. The number of wavelengths that pass a fixed point in one second is the frequency of the wave. Frequency is measured in a unit called hertz (Hz). A frequency of 1 Hz indicates that one wavelength is passing a point each second. The frequency can be found using the following equation:

frequency = number of wavelengths/1 second

The velocity of a wave depends upon the material through which the wave passes. The velocity of a wave is equal to its wavelength times its frequency. A wave's velocity is expressed in the same units as any measurement of velocity—meters per second (m/s).

velocity = wavelength × frequency

Strategy

You will identify the crest, trough, and amplitude of a wave.
You will determine the wavelength and frequency of a wave.
You will calculate the velocity of a wave.

Materials 🥽

instant developing camera
meterstick
20 pieces of colored yarn
rope, about 5 m long
    or
coiled spring toy

Procedure

Part A—Frequency of a Wave

1. Safety goggles should be worn throughout the experiment. Tie the pieces of yarn to the rope at 0.5 m intervals. Use the meterstick to measure the distances.

2. Tie one end of the rope to an immovable object, such as a table leg. Pull the rope so it does not sag.

3. Make waves in the rope by moving the free end up and down. Continue to move the rope at a steady rate. Observe the crests, troughs, and amplitude of the waves.

Laboratory Activity 1 (continued)

4. Continue making waves by moving the rope at a constant rate. Observe a particular piece of yarn. Count the number of wavelengths that you produce during a period of 30 seconds. Record this value in Table 1 as wave motion A.

5. Slow the rate at which you are moving the rope. Predict what will happen to the frequency. Count the number of wavelengths produced in 30 seconds while maintaining this constant slower rate. Record this value in Table 1 as wave motion B.

6. Repeat the procedure in step 4 moving the rope at a faster rate. Maintain this constant rate for 30 seconds. Record this value in Table 1 as wave motion C.

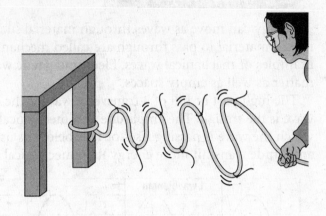

Part B—Velocity of a Wave

1. Using the same rope setup as in Part A, have a classmate move the rope with a constant motion. Record the number of wavelengths produced in 30 seconds in Table 2 as wave motion A. Photograph the entire length of the moving rope using the instant developing camera. Rest the camera on a table to keep it still.

2. Have your classmate increase the motion of the rope and take another photograph. Predict what will happen to the wavelength. Again count the number of wavelengths produced in 30 seconds, and record these values in Table 2 as wave motion B.

3. Observe the developed photographs. For each photograph, use the yarn markers to determine the length of one wavelength. Record these values in Table 2. You may tape the photographs to the last page of this Laboratory Activity.

4. Calculate the frequency of each of the three waves produced in Part A. Use the equation for the frequency found in the introduction. Record the values of the frequencies in Table 1.

5. Calculate the frequencies of the two waves produced in Part B. Record these values in Table 2.

6. Calculate the velocities of the two waves using the values of the wavelengths and frequencies in Table 2. Use the equation for velocity of a wave found in the introduction. Record the values of the velocities in Table 2.

Laboratory Activity 1 (continued)

Data and Observations

Part A—Frequency of a Wave

Wave motion	Number of waves in 30 s	Frequency (Hz)
A		
B		
C		

Part B—Velocity of a Wave

Wave motion	Number of waves in 30 s	Frequency (Hz)	Wavelength (*m*)	Velocity (*m/s*)
A				
B				

Questions and Conclusions

1. As you increased the motion of the rope, what happened to the frequency of the waves?

_____

_____

2. As the frequency of the waves increased, what happened to the wavelength?

_____

_____

3. As the frequency of the waves increased, what happened to the velocity of the waves?

_____

_____

4. Does your data indicate that the velocity of a wave is dependent or independent of its frequency? Explain.

_____

_____

_____

Strategy Check

_____ Can you identify the crest, trough, and amplitude of a wave?

_____ Can you determine the wavelength and frequency of a wave?

_____ Can you calculate the velocity of a wave?

Laboratory Activity 1 (continued)

Attach photographs here.

Waves in Motion

Have you ever tossed a pebble into a puddle and watched the ripples? The ripples are actually small water waves. Have you wondered what affects those ripples? In this Lab Activity, you will look at ripples and how they behave.

Strategy

You will observe wave phenomena in a ripple tank.

Materials

ripple tank with light source and
   bottom screen
ripple bar
*3/4-in dowel, about 5 cm shorter than
   ripple tank*
paraffin block

dropper
glass plate, about 1/4 the area of the
   ripple tank
rubber stoppers cut to 1.5 cm high (2)
Alternate materials

Procedure

1. Turn on the light of the ripple tank. Allow the water to come to rest. Touch your finger once to the water surface to produce a wave. On the screen at the base of the tank, observe the shape of the wave. Does the speed of the wave seem to be the same in all directions? Record your observations in the table in the Data and Observations section.

2. Place the ripple bar in the water. Allow the water to come to rest. Using the flat of your hand to touch *only* the ripple bar, roll the ripple bar forward 1 cm. Observe the wave you produce. Record your observations in the table in the Data and Observations section. NOTE: Be careful to touch only the ripple bar when generating waves, do not touch the water with your hand.

3. Place a paraffin block in the tank parallel and closer to the deep end of the ripple tank. Orient the ripple bar to be parallel to the long edge of the paraffin block. Allow the water to come to rest. Use the flat of your hand to roll the ripple bar forward 1 cm, generating a wave that strikes the paraffin block barrier straight on. Observe what happens to the wave when it reaches the barrier. How does the wave move after it strikes the barrier? Record your observations in the Data and Observations section.

4. Reposition the paraffin block so that it is not aligned with the edges of the tank. This will change the angle at which the wave strikes it. Position the ripple bar so that it is parallel and closer to the shallow edge of the tank. After the water has come to a rest, move the ripple bar forward 1 cm with the flat of your hand. Observe the shape of the waves that reflect off the paraffin block. Record your observations. Remove the ripple bar from the water.

5. Allow the water to come to rest. Use the dropper to drop one drop of water onto the water surface. Observe the circular wave shape. Take note of how the wave reflects from the paraffin block and the point from which the reflected wave appears to originate. Record your observations in the Data and Observations section.

6. Place a paraffin barrier on one side of the tank, halfway between the shallow end and the deep end of the tank. Place the ripple bar parallel and closer to the shallow end. Again use a ripple bar to produce a straight wave. See step 3. Observe the part of the wave that strikes the barrier as well as the part that passes by it. Record your observations in the table.

Laboratory Activity 2 (continued)

7. Support a piece of glass with rubber stoppers so that the glass is in the shallow end of the tank 1.5 cm from the bottom of the tank and its top is just covered with water. Position the glass so that the edges of the glass are parallel to the edges of the tank. Place the ripple bar in the deep end of the tank, parallel to the edge. Allow the water to come to rest. Then move the ripple bar 1 cm to create a wave. Observe what happens as the waves pass from the deep to the shallow end of the tank. Record your observations in the Data and Observations section.

8. Turn the glass so that its edges are no longer parallel to the edges of the ripple tank. Allow the water to come to rest, and then repeat step 7. Observe the shape of the waves that pass over the glass and that pass around the glass. Also note the speed of these waves. Record your observations.

Data and Observations

Step	Question	Observation
1	What is the shape of the wave?	1.
1	Is the speed of the wave the same in all directions?	2.
2	What is the shape of the wave?	3.
3	What happens to the wave at the barrier?	4.
3	What is the direction of the wave after it strikes the barrier?	5.
4	What is the shape of the reflected wave?	6.
5	How does the wave reflect from the paraffin block?	7.
5	From what point does the reflected wave appear to originate?	8.
6	What happens to the wave that hits the block?	9.
6	What happens to the wave that does not hit the block?	10.
7	What happens as waves pass from deep to shallow water?	11.
8	What is the shape of the wave that passes over the glass?	12.
8	What is the shape of the wave that does not pass over the glass?	13.
8	How do the speed of the two different waves compare?	14.

Laboratory Activity 2 (continued)

Questions and Conclusions

1. What is the shape of a wave produced at one point, such as with a drop of water or your fingertip?

_____

2. What does a wave do when it hits a paraffin barrier?

_____

3. Does a circular wave remain circular when it is reflected? Explain why this happens.

_____

_____

4. What happens to waves as they move into shallower water?

_____

_____

Strategy Check

_____ Can you identify behavior of waves?

Sound Waves and Pitch

Sounds are produced and transmitted by vibrating matter. You hear the buzz of a fly because its wings vibrate, the air vibrates, and your eardrum vibrates. The sound of a drum is produced when the drumhead vibrates up and down, the air vibrates, and your eardrum vibrates. Sound is a compressional wave. In a compressional wave, matter vibrates in the same direction as the wave travels. For you to hear a sound, a sound source must produce a compressional wave in matter, such as air. The air transmits the compressional wave to your eardrum, which vibrates in response to the compressional wave.

Compressional waves can be described by amplitude, wavelength, and frequency— the same as transverse waves. The pitch of a sound is related to the frequency of a compressional wave. You are familiar with high pitches and low pitches in music, but people are also able to hear a range of pitches beyond that of musical sounds. People can hear sounds with frequencies between 20 and 20,000 Hz.

Strategy

You will demonstrate that sound is produced by vibrations of matter.
You will vary the pitch of vibrating objects.

Materials

4 rubber bands of different widths but equal lengths
cardboard box, such as a shoe box or cigar box

Safety Precautions 🥽

Safety goggles should be worn throughout the experiment.

Procedure

1. Stretch the four rubber bands around a box as shown in Figure 1.

Figure 1

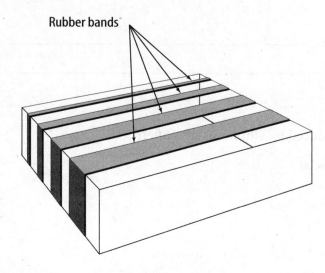

Rubber bands

2. Pluck the first rubber band, allowing it to vibrate. Listen to the pitch of the vibrating rubber band. Predict how the pitches of the other rubber bands will compare with this pitch. Record your prediction in the Data and Observations section. Pluck the remaining rubber bands. Record your observations about the variation in pitch.

3. Remove three rubber bands from the box. Hold the remaining rubber band tightly in the middle with one hand. Pluck it with the other. Move your hand up and down the rubber band to increase or decrease the length of the rubber band that can vibrate. Predict how the pitch will change as you change the length of the vibrating rubber band. Pluck the rubber band for each new length and record your observations of the length of the vibrating rubber band and pitch.

Laboratory Activity 1 (continued)

Data and Observations

1. Prediction of variation in pitch of sounds produced by rubber bands of different widths:

_____

_____

2. Observation of changes in pitch with varying thickness of rubber bands:

_____

_____

_____

3. Observation of changes in pitch with varying length of the rubber band:

_____

_____

_____

_____

Questions and Conclusions

1. How does length affect the pitch of sound produced by a vibrating object?

_____

_____

2. How does the width of a rubber band affect its frequency of vibration?

_____

_____

3. Based on your results, how would you expect the pitch of sound produced by a vibrating string to be affected by the length of the string?

_____

_____

Strategy Check

_____ Can you demonstrate that sound is produced by vibrations of matter?

_____ Can you vary the pitch of vibrating objects?

LAB 2 Laboratory Activity
Musical Instruments

Musical instruments have been made and used for thousands of years by different cultures around the world. String, brass, woodwind, and percussion instruments all produce their own distinctive musical sounds. In this activity, you can make and compare the sounds made by several simple instruments.

Strategy

You will construct simple musical instruments.
You will compare and contrast the sounds made by these instruments.
You will classify the instruments according to their type.

Materials

block of wood, 15 cm × 10 cm × 5 cm
wire coat hanger
wire cutters
wire staples
hammer
6 beakers of the same size
wooden spoon
water
shoe box or tissue box
2 pieces of wood, 1 cm × 1 cm × 15 cm
5 rubber bands of varying lengths and
    thicknesses
6 nails of varying lengths, 5 cm to 20 cm

meterstick
string, 90 cm
scissors
metal spoon
2 plastic soda bottles with lids
dried peas, small pebbles, uncooked rice, or
    paper clips
plastic trash bag
string
tape
empty containers such as margarine tubs,
    plastic bowls, or cardboard tubes

Procedure

Part A—Twanger

1. Use wire cutters to cut a coat hanger into four or five pieces of different lengths. The lengths of the pieces should vary from 8 cm to 20 cm.

2. Use wire staples and a hammer to attach the lengths of wire to the wooden block, as shown in Figure 1.

Figure 1

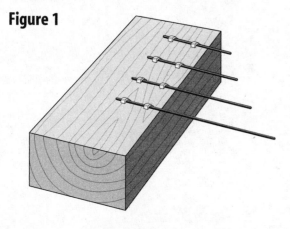

3. Pluck the wires with your thumb or a pen. In the data table in the Data and Observations section, describe the sounds and pitches produced by the various pieces of wire.

Part B—Xylophone

4. Set up six beakers of the same size in a row.

5. Leave the first beaker empty. Add increasing amounts of water to each of the remaining five beakers. The third beaker should be about half full and the last beaker should be almost full.

6. Tap the side of each beaker gently with a wooden spoon. Describe the sounds and pitches produced by each of the beakers.

Laboratory Activity 2 (continued)

Part C—Guitar

7. Stretch the rubber bands around the box lengthwise.

8. With a partner's help, slide one piece of wood under the rubber bands at one end of the box. Slide the other piece of wood under the rubber bands at the other end of the box. Your completed guitar should look like Figure 2.

9. Pluck the rubber bands with your fingers. Describe the sounds and pitches produced by each of the rubber bands.

Figure 2

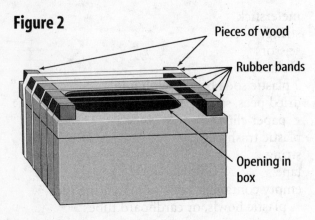

Pieces of wood

Rubber bands

Opening in box

Part D—Nail Chimes

10. Cut a piece of string into 15-cm pieces. Tie each piece of string around the meterstick, leaving a long end hanging down.

11. Tie the hanging end of each string around the head of a nail. Arrange the nails from shortest to longest.

12. Suspend the meterstick between two chairs or tables, as shown in Figure 3. Be sure the nails don't touch each other.

13. Use a metal spoon to tap the nails. Describe the sounds and pitches produced by each of the nails.

Figure 3

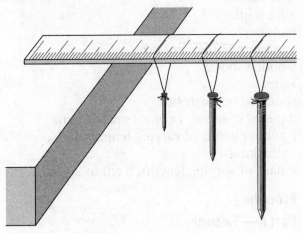

Laboratory Activity 2 (continued)

Part E—Shakers

14. Place a small amount of dried peas, small pebbles, uncooked rice, or paper clips into one plastic bottle. Screw the cap on the bottle.

15. Place a small amount of another material in a second plastic bottle and screw on the cap.

16. Shake each bottle or tap it against your hand. Describe the sounds made by each shaker.

Part F—Drums

17. Cut down the sides of the plastic bag to make one large sheet.

18. Place one of your containers open side down on the plastic sheet. Cut around the container, leaving about an extra 10 cm all around the container.

19. With a partner, stretch the plastic tightly over the top of the container. Use string and tape to hold the plastic in place, as shown in Figure 4.

20. Repeat steps 18 and 19 with a container of a different size.

21. Hit the top of your drums lightly with your fingers or a pencil. Describe the sounds made by each drum.

Figure 4

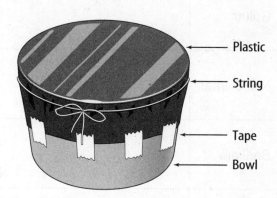

Plastic
String
Tape
Bowl

Laboratory Activity 2 (continued)

Data and Observations

Instrument	Sounds and pitches
Twanger	
Xylophone	
Guitar	
Nail chimes	
Shakers	
Drums	

Questions and Conclusions

1. Which instruments were able to produce sounds of different pitches?

_____

2. What caused the different pitches of sounds in each of those instruments?

_____

3. Classify each of the instruments you made by type.

_____

_____

4. How does the length of a piece of wire or a nail affect its frequency of vibration?

_____

Strategy Check

_____ Can you construct simple musical instruments?

_____ Can you compare and contrast the sounds made by these instruments?

_____ Can you classify the instruments according to their type?

Observing the Electromagnetic Spectrum

Electromagnetic waves are produced by electric charges that move or vibrate back and forth. The frequency of the electromagnetic wave is the same as the frequency at which the charge vibrates. Except for their frequency and wavelength, all electromagnetic waves are the same and travel at the same speed, 300,000 km/s in a vacuum. The electromagnetic spectrum is used to classify electromagnetic waves according to their frequency and wavelength. Humans are only able to see visible light, which is one small portion of the spectrum. In this activity, you will create a model of the infrared, visible, and ultraviolet portions of the electromagnetic spectrum. The model you create will be made to scale based on wavelength.

Strategy

You will create a scale model of portions of the electromagnetic spectrum.
You will demonstrate that visible light makes up a very small portion of the electromagnetic spectrum.

Materials

calculator
meterstick or metric ruler (marked in millimeters)
scissors
one piece of paper in each of the following colors: red, orange, yellow, green, blue, violet, white, and black
black marker
cellophane tape
flashlight
prism

Procedure

1. The wavelengths for the visible, infrared, and ultraviolet portions of the spectrum are represented in meters in the data table. Complete a metric conversion calculation to find the length of the waves in nanometers. One nanometer is 1×10^{-9} of a meter.

   Use the following conversion factors to convert the wavelengths to nanometers:
   10^{-6}M = 1,000 nanometers;
   10^{-7}M = 100 nanometers; and
   10^{-8}M = 10 nanometers.

2. The scale that will be used to build the model of the spectrum is 1 nanometer equals 1 millimeter. Therefore, if a wavelength is x nanometers, the model for that wavelength should measure x millimeters. Record your calculations for scale wavelength in millimeters in the table in the Data and Observations section.

3. Work together as a class on the metric conversion calculation for red light. It is good to begin with red light rather than infrared, which is listed first in the data table, because the length of the scale model for infrared light is significantly longer than the scale models of any of the visible light colors.

4. Fill in the scale length in the millimeters column in your data table for red light. This column should always be the same as the final answer for wavelength in nanometers.

5. Use the colored paper to represent the different colors in the visible spectrum. Red paper will be used for the wavelength of red light, orange paper for orange light, and so on. White paper will represent infrared, and black paper will represent ultraviolet.

Laboratory Activity 1 (continued)

6. Cut a strip of red paper that is 2.5 cm wide and the same length as the number you have written in your column for scale length in millimeters.

7. Once you have a strip of red paper that is 750 mm (75 cm) long, mark the actual wavelength of red light, 7.5×10^{-7} m, on the strip.

8. Complete a metric conversion calculation and cut strips for each of the electromagnetic waves represented in the data table. When you have finished, you should have eight strips of paper of different lengths and colors in your model.

9. Align your strips horizontally, directly underneath each other, with the longest strip (which should be infrared) on top and the shortest strip (which should be ultraviolet) on the bottom. Tape all the strips together to make one large sheet.

10. Shine the flashlight through the prism in order to see the visible spectrum you have just modeled.

Data and Observations

Wave	Actual wavelength in meters	Calculation	Actual wavelength in nanometers	Scale wavelength in millimeters
1. Infrared	1×10^{-6}			
2. Red	7.5×10^{-7}			
3. Orange	6.25×10^{-7}			
4. Yellow	5.75×10^{-7}			
5. Green	5.25×10^{-7}			
6. Blue	4.5×10^{-7}			
7. Violet	4×10^{-7}			
8. Ultraviolet	3×10^{-8}			

Questions and Conclusions

1. What colors did you observe from the prism? List them in order from top to bottom.

2. Why do we use the scale of 1 nanometer equals 1 millimeter?

Laboratory Activity 1 (continued)

3. How many times longer is the wavelength of the infrared wave than the wavelength of the ultraviolet wave?

_____

_____

_____

_____

4. A radio wave is approximately 3 meters long. If we were to make this wave part of our model, how long would the strip of paper representing the radio wave have to be?

_____

_____

5. Why didn't you include the whole electromagnetic spectrum in your model?

_____

_____

_____

_____

Strategy Check

_____ Can you compare the wavelength sizes of the electromagnetic spectrum?

_____ Can you demonstrate that visible light is a small portion of the electromagnetic spectrum?

Catching the Wave

When you listen to your radio, you are hearing the information that is carried by electromagnetic waves. Many electronic and electrical devices produce low frequency electromagnetic waves. In this lab, you will detect these waves.

Strategy

You will detect electromagnetic waves.
You will determine what produces electromagnetic waves.

Materials

9-V battery
mini audio amplifier
telephone pickup coil
strong permanent magnet
string
variety of appliances: calculators, watches, radios, computers, lights, burning candles, vacuum cleaners

Procedure

1. Put the 9-V battery into the amplifier and plug in the telephone pickup coil.
2. Place the strong permanent magnet on the table. Slowly bring the pickup coil near the magnet.
3. Hold the pickup coil still. Attach the magnet to a string so that it hangs just above the coil and allow it to swing.

Listen for the electromagnetic waves. Record your observations in the table in the Data and Observations section.

4. Use the pickup coil to try to pick up signals from televisions, computers, lights, burning candles, appliances, watches, etc. Record your observations.

Data and Observations

Appliance	Sound
Pickup coil and amplifier	

Laboratory Activity 2 (continued)

Questions and Conclusions

1. What causes sound waves?

_____

_____

_____

_____

2. Does the sound change when two items are close to the coil?

_____

_____

_____

_____

_____

3. Why can you hear the electromagnetic waves from some objects that are not electrical appliances?

_____

_____

_____

_____

4. How can you predict which objects will produce an electromagnetic wave and which will not?

_____

_____

_____

_____

_____

Laboratory Activity 2 (continued)

5. Some people think that electromagnetic waves are harmful to our health. Do you think we could get rid of these waves?

_____

_____

_____

_____

_____

Strategy Check

_____ Can you detect electromagnetic waves?

_____ Can you predict if an item will produce electromagnetic waves?

Producing a Spectrum

Chapter 12

1 Laboratory Activity

Each color of light has a particular wavelength. The colors that make up white light can be separated into individual colored bands called a spectrum. A spectrum can be produced by refraction or interference.

When light passes from one substance into another, its speed changes. If a ray of light passes from one substance into another at an angle, its direction also changes. The change in speed and possible change in direction of light as it enters a substance is called refraction. Refraction of light can be seen with a prism. As light enters a prism, each wavelength is bent a different amount. Thus, the wavelengths are separated into a spectrum.

When you see colors in soap bubbles or in drops of oil on wet pavements, you are observing the interference of light rays. Some of the light striking the outer surface of a thin film, such as a soap bubble, is reflected to your eyes. Some of the light passing through the bubble film is reflected from the inner surface of the film to your eyes. The rays from the inner surface travel a slightly longer path than the rays reflected from the outer surface. The waves do not arrive at your eyes together. They are out of phase. Your eyes may receive the crest of one wave along with the trough of another wave.

Waves out of phase cancel each other and no color is produced. Waves that are out of phase undergo destructive interference. Other areas of the thin film reflect light rays that are in phase, and you see bands of color. The colors you see are due to constructive interference. The interference of light reflecting from the other two surfaces of a thin film creates bands of different colors. These bands change position if the viewing angle changes or the film changes thickness.

Strategy

You will describe the spectrum made by a prism.
You will describe an interference pattern.
You will explain how a spectrum can be produced by refraction of light and by interference.

Materials

Part A
projector or other
   light source
prism
tape
white paper

Part B
bowl
water
index card
clear nail polish
scissors

Procedure

Part A—Refraction

1. Darken the room. Your teacher will set up a projector or other light source in the room. **CAUTION:** *Do not look directly into the light source.* Hold a prism in the beam of the projector so the light strikes one of the three rectangular sides. Rotate the prism by holding the triangular ends so a pattern of colors is produced on the wall. Tape a piece of white paper to the wall where the spectrum appears.

2. Observe the spectrum and the order of the colors. Write in the names of the colors, in the order you see them, on Figure 3 in the Data and Observations section.

Laboratory Activity 1 (continued)

Part B—Interference

1. Fill the bowl with water.
2. Cut two holes, each about 1 cm in diameter, in the index card.
3. Place two separate drops of nail polish on the surface of the water. The polish will harden to a film. Position one hole of the index card under one of the films of nail polish and carefully lift the film from the water. Repeat, using the second hole and the second film. Hold the card vertically and allow the films to dry in place.
4. Hold the index card away from the light, as shown in Figure 2. Look down at the films. If you change the position of your head while viewing the card, you should observe a spectrum on each film.
5. Describe and make a drawing of the color patterns you observe in the Data and Observations section.

Figure 1

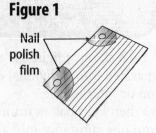

Nail polish film

Figure 2

Data and Observations

Part A

1. Label the colors of the spectrum you observed.

Figure 3

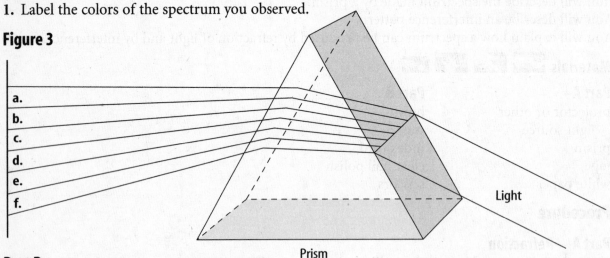

a.

b.

c.

d.

e.

f.

Prism

Light

Part B

1. Description of color patterns on thin film:

_____

_____

2. Drawing of the interference pattern on the thin film:

Laboratory Activity 1 (continued)

Questions and Conclusions

1. What is refraction of light?

_____

_____

_____

2. What is destructive interference? What is constructive interference?

_____

_____

_____

3. Label the surfaces of the cross-sectional diagram of the thin film shown in Figure 4. Show the two things that happen to a ray of light that strikes the outer surface. Show what happens to a ray of light that continues and strikes the inner surface. Label each instance of reflection and refraction that occurs when light strikes a thin film.

Figure 4

4. Compare and contrast the production of a spectrum from a prism and from a thin film.

_____

_____

_____

_____

Laboratory Activity 1 (continued)

5. Label each figure below as an example of reflection, refraction, or interference.

Figure 5

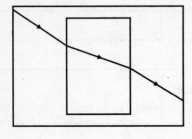
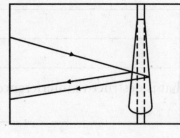

a. _____   b. _____   c. _____

Strategy Check

_____ Can you describe the spectrum made by a prism?

_____ Can you describe an interference pattern?

_____ Can you explain how a spectrum can be produced by refraction of light and by interference?

Light Intensity

LAB 2 Laboratory Activity

Have you ever noticed how the brightness of the light from a flashlight changes as you move closer to or farther away from it? Likewise, have you ever noticed how the strength of the signals from a radio station fades on a car radio as you move away from the transmitting tower? Both light and radio signals are electromagnetic waves. These two examples seem to suggest that the intensity of energy and distance are related. What is the relationship between light intensity and distance? Is there also a relationship between light intensity and direction?

In this experiment you will use a photo resistor, a device that changes its resistance to an electric current according to the intensity of the light hitting it. The resistance is measured in a unit called an ohm (Ω). Photo resistors are often used in burglar alarm systems. A beam of light shines on the photo resistor. If anyone or anything passes through the beam, the intensity of the light is changed. Because the photo resistor is in an electric circuit, the current in the circuit changes and this causes an alarm to sound.

Strategy

You will measure the effect of distance on light intensity.
You will measure the effect of direction on light intensity.
You will interpret graphs relating light intensity, distance, and direction.

Materials

photo resistor
pencil
tape
ring stand
meterstick
black tape

multimeter or
    ohmmeter
25-W lightbulb
    and lamp socket
utility clamp
colored pencils

Procedure

1. In the Data and Observations section, write hypotheses explaining the relationships between light intensity and distance and between light intensity and direction.

2. Mount the photo resistor on a pencil with tape. See Figure 1.

3. Lay the meterstick on a flat, hard surface. Place small pieces of black tape at 0.10 m intervals along the meterstick.

4. Set the lightbulb and socket on a smooth, flat surface.

5. Clamp the meterstick to the ring stand with the utility clamp. Arrange the meterstick so that the lightbulb is at the 0.00 m marker. See Figure 2.

Figure 1

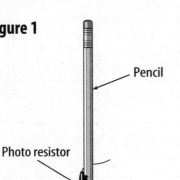

Pencil

Photo resistor

Tape

Figure 2

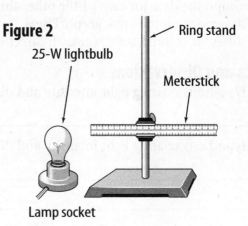

Ring stand

25-W lightbulb

Meterstick

Lamp socket

Laboratory Activity 2 (continued)

Figure 3

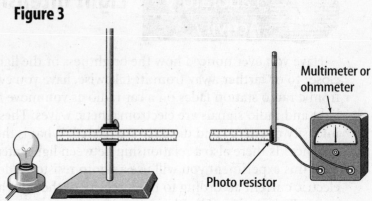

Multimeter or ohmmeter

Photo resistor

6. Attach the wires of the photo resistor to the multimeter or ohmmeter. If using a multimeter, set the meter to measure resistance and attach the wires to the appropriate terminals. Darken the room before any measurements are taken.

7. Turn off the bulb and place the photo resistor at the 1.00 m marker. See Figure 3.

8. Measure the resistance using the multimeter or ohmmeter. Record the value in Table 1 in the column marked *East*.

9. Move the photo resistor to the 0.90 m marker. Record the value in the same column of the data table.

10. Continue advancing the photo resistor to each marker. Record the meter reading at each position. The last reading should be taken at the 0.10 m marker.

Figure 4

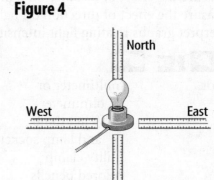

North

West

East

South

11. Assume that the meterstick was oriented with the 1.0 m marker pointing to the East. Repeat the procedure for each of the three remaining directions shown in Figure 4.

12. Use Graph 1 to graph your data. Place the distance values on the *x*-axis and the resistance values on the *y*-axis. Label the *x*-axis *Distance from light source (m)* and the *y*-axis *Resistance (Ω)*.

13. Graph the data for each of the other three directions on the same graph. Use a different colored pencil for each direction.

Data and Observations

1. Hypothesis relating light intensity and distance:

_____

2. Hypothesis relating light intensity and direction:

_____

_____

_____

Laboratory Activity 2 (continued)

Table 1

Distance (m)	Resistance (Ω)			
	East	West	North	South
1.00				
0.90				
0.80				
0.70				
0.60				
0.50				
0.40				
0.30				
0.20				
0.10				

Graph 1

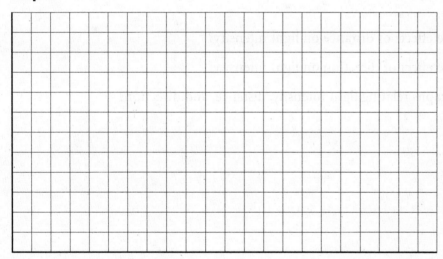

Questions and Conclusions

1. Look at the graph. Describe how the resistance and distance are related.

_____

_____

2. How are light intensity and distance related?

_____

Laboratory Activity 2 (continued)

3. What does the graph indicate about the relationship between intensity of light and direction?

_____

_____

_____

4. Why was it necessary to darken the room before doing this experiment?

_____

_____

_____

5. Do the results of this experiment support your original hypotheses?

_____

6. Light from the Sun travels to Earth from a distance of almost 150 million km. If Earth were farther away from the Sun, what effects would be felt on Earth's surface? What effects would be felt if Earth were closer to the Sun?

_____

_____

_____

Strategy Check

_____ Can you measure the effects of distance on light intensity?

_____ Can you measure the effect of direction on light intensity?

_____ Can you interpret graphs relating light intensity, distance, and direction?

Reflection of Light

Chapter 13

Light travels in straight lines. When a light ray strikes a smooth surface, such as polished metal or still water, it is reflected. The angle between the incoming ray, the incident ray, and the normal line is called the angle of incidence. The normal line is a line forming a right angle with the reflecting surface as shown in Figure 1. The angle between the reflected ray and the normal line is called the angle of reflection.

Rough or irregular surfaces reflect parallel light rays in all directions. Because light is reflected from rough surfaces in all directions, these surfaces cannot be used to produce sharp images.

Figure 1

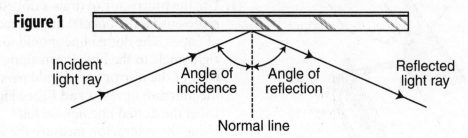

Strategy

You will observe that light travels in straight lines.

You will identify the angles of incidence and reflection of reflected light.

You will describe the relationship between the angle of incidence and the angle of reflection.

Materials

white paper (3 sheets)   book
flashlight or projector   plane mirror
masking tape   comb
pen or pencil   protractor

Procedure

1. Use masking tape to attach one sheet of white paper to the cover of the book. Tape the comb to the edge of the book. The teeth of the comb should extend above the edge of the book as shown in Figure 2.

2. Darken the room. Shine the flashlight through the comb onto the paper from as far away as possible. Support the flashlight on a table or books. Observe the rays of light on the paper. Record your observations in the Data and Observations section.

Figure 2

Teeth extend above edge of book.

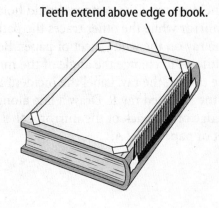

Laboratory Activity 1 (continued)

3. Stand the plane mirror at a right angle to the surface of the book cover. Position the mirror about two-thirds of the width of the book away from the comb. Adjust the mirror so that the light rays hit it at right angles. See Figure 3.

Figure 3

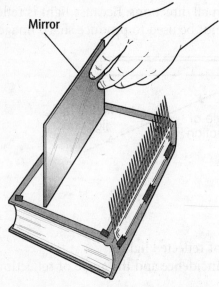

Mirror

4. Rotate the mirror so that the light rays strike it at various angles of incidence. As you turn the mirror, observe the reflected rays of light. Form a hypothesis relating the incident and reflected rays and write it in the Data and Observations section.

5. With the mirror turned so the incident rays strike it at an angle of about 30°, study a single incident ray. One partner should hold the mirror while the other traces the path of the ray on the white sheet of paper. Be careful not to change the angle of the mirror while tracing the ray. Label the incident ray *I* and the reflected ray *R*. Draw a line along the edge of the back of the mirror. Label the sheet of paper *Trial A*.

6. Repeat step 5 using a new sheet of paper on the book. Hold the mirror at a greater angle and trace the ray and the edge of the back of the mirror. Label this sheet *Trial B*. Repeat step 5 for a third time and label the sheet of paper *Trial C*.

7. After analyzing the ray tracings, attach them to page 11.

Analysis

1. Use the protractor to draw a dotted line representing the normal line on each sheet of paper. The dotted line should form a right angle to the line drawn along the back edge of the mirror and should pass through the junction of rays *I* and *R*. See Figure 4. Label the dotted line *normal line*.

2. Using the protractor, measure the angle between the normal line (dotted line) and the tracing of the incident ray (*I*) for Trial A. Record this value in Table 1. Measure the angle between the normal line and the tracing of the reflected ray (*R*), and record this value in the data table. Measure and record the angles for Trials B and C in the same way.

Figure 4

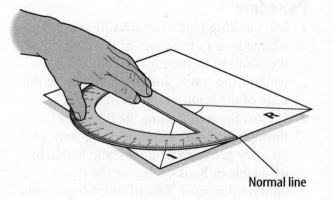

Normal line

Laboratory Activity 1 (continued)

Data and Observations

Observation of light rays in step 2 of the procedure:

_____

Hypothesis:

_____

_____

_____

Table 1

Trial	Angle of incidence	Angle of reflection
A		
B		
C		

Attach ray tracings here.

Laboratory Activity 1 (continued)

Questions and Conclusions

1. Explain how your observations of light passing between the teeth of a comb support the statement that light travels in straight lines.

_____

_____

_____

_____

2. Why did you mark the position of the *back* edge of the plane mirror on your ray tracings?

_____

_____

_____

_____

3. As you increased the angle of incidence, what happened to the angle of reflection?

_____

_____

4. Explain the relationship between the angle of incidence and the angle of reflection.

_____

_____

_____

_____

Strategy Check

_____ Can you identify the angles of incidence and reflection of reflected light?

_____ Can you explain the relationship between the angles of incidence and reflection?

LAB 2 Laboratory Activity

Magnifying Power

Parallel rays of light passing through a convex lens are refracted toward a single point called the focal point. Rays passing through the center of the lens are not bent at all. Rays passing through the edges of the lens bend sharply. The distance between the focal point and the midpoint of the lens is the focal length. As you can see in Figure 1, the curvature of a lens determines the position of its focal point. A thin lens has only a slight curvature. It has a long focal length because the focal point is far from the lens. A thicker lens has a greater curvature. It has a shorter focal length because its focal point is closer to the center of the lens.

Figure 1

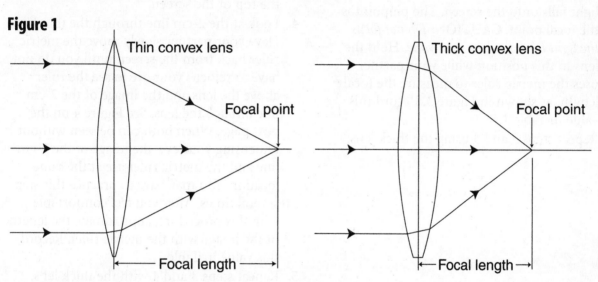

Thin convex lens — Focal point — Focal length

Thick convex lens — Focal point — Focal length

You have probably used a magnifying glass to make objects appear larger. A magnifying glass is a convex lens. The magnifying power of a convex lens indicates how many times larger the image is compared to the object. A lens with a magnifying power of 3× indicates that the image of a line 1 cm long will appear to be 3 cm long when viewed through the lens.

The magnifying power of a convex lens can be calculated if the focal length of the lens is known. How is the magnifying power of a lens related to its focal length?

Strategy

You will measure the focal length of two lenses.
You will predict the magnifying power of each lens.
You will determine the magnifying power of each lens.

Materials

book
large white index card
masking tape
thick convex lens
thin convex lens
metric ruler

Laboratory Activity 2 (continued)

Procedure

Part A—Measuring the Focal Length of a Convex Lens

1. Make a screen by taping the large white index card to a book as shown in Figure 2.
2. Using the thin lens, focus the light from a bright light source onto the screen. Direct sunlight is best, but a lamp can be used.
3. Adjust the lens so that a small pinpoint of light falls onto the screen. The pinpoint is the focal point. **CAUTION:** *Do not focus the light from the lens at anyone.* Hold the lens in this position while your partner uses the metric ruler to measure the focal length, as shown in Figure 3. Record this value in Table 1.
4. Repeat steps 2 and 3 using the thick lens.

Part B—Determining the Magnifying Power of a Lens

1. Use the metric ruler to draw a 2.0-cm horizontal line about 0.5 cm below the top of the screen.
2. Place the screen about 25 cm away.
3. Have your partner hold the metric ruler slightly behind and parallel to the screen. The bottom of the ruler should align with the top of the screen.
4. Look at the 2-cm line through the thin lens. Have your partner slowly move the metric ruler back from the screen until you do not have to refocus your eye to see the ruler above the lens and the image of the 2-cm line *through* the lens. See Figure 4 on the next page. When both can be seen without refocusing your eye, the image of the 2-cm line and the metric ruler are at the same location. You may have to practice this step several times. After you feel comfortable with this procedure, approximate the length of the image with the metric ruler. Record this value in Table 1.
5. Repeat steps 3 and 4 with the thick lens.

Figure 2

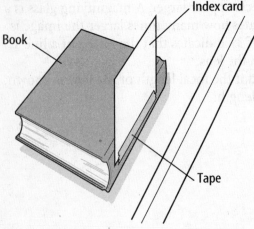

Figure 3

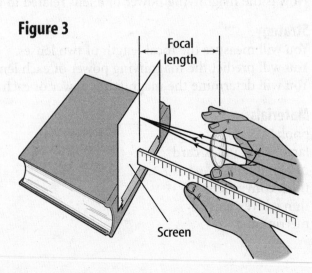

Laboratory Activity 2 (continued)

Figure 3

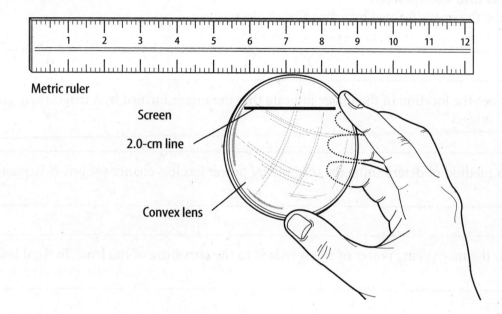

Metric ruler

Screen

2.0-cm line

Convex lens

Data and Observations

Table 1

Lens	Focal length (cm)	Length of image (cm)
thin		
thick		

Table 2

Lens	Magnifying power	
	Method 1	Method 2
thin		
thick		

1. Calculate the magnifying power of each lens by finding the ratio of the length of the image and the 2-cm line. Record these values under Method 1 in Table 2.

2. The magnifying power of a lens can also be calculated using the following equation.

$$magnifying\ power = \frac{focal\ length}{25\ cm - focal\ length}$$

Use this equation to calculate the magnifying factor of each lens. Record these values under Method 2 in Table 2.

Laboratory Activity 2 (continued)

Questions and Conclusions

1. Describe the image formed by a magnifying glass.

   _____

   _____

2. How does the location of the image indicate that the image formed by a magnifying glass is a virtual image?

   _____

3. Which method of determining the magnifying power has less chance for error? Explain.

   _____

   _____

4. How is the magnifying power of a lens related to the curvature of the lens? To focal length?

   _____

   _____

5. Can a drop of water resting on a surface act as a magnifying glass? Explain.

   _____

   _____

   _____

6. A student draws a 1-cm × 2-cm rectangle on a piece of paper. What will be the area of the image of the rectangle if the student observes the drawing through a 2× hand lens?

   _____

Strategy Check

_____ Can you measure the focal length of a concave or convex lens?

_____ Can you determine the magnifying power of a lens?

Density of a Liquid

All matter has these two properties – mass and volume. Mass is a measure of the amount of matter. Volume is a measure of the space that the matter occupies. Both mass and volume can be measured using metric units. The standard unit of mass in the SI system is the kilogram (kg). To measure smaller masses, the gram (g) is often used. In the metric system, the volume of a liquid is measured in liters (L) or milliliters (mL). Density is a measure of the amount of matter in a given volume of space. Density may be calculated using the following equation.

$$\text{density} = \frac{\text{mass}}{\text{volume}}$$

Density is a physical property of a liquid. By measuring the mass and volume of a sample of a liquid, the liquid's density can be determined. The density of a liquid is expressed as grams per milliliter (g/mL). For example, the density of distilled water is 1.00 g/mL.

Strategy

You will determine the capacity of a pipette.
You will measure the masses of several liquids.
You will calculate the densities of the liquids.
You will compare the densities of the liquids with that of water.

Materials 🚫 🥽 🧪 ✋

4 plastic pipettes
metric balance
distilled water
4 small plastic cups
ethanol
corn oil
corn syrup

Procedure

Part A—Determining the Capacity of a Pipette

1. Measure the mass of an empty pipette using the metric balance. Record the mass in the Data and Observations section.
2. Completely fill the bulb of the pipette with distilled water. This can be done as follows:
   a. Pour distilled water into a small plastic cup until it is half full.
   b. Squeeze the bulb of the pipette and insert the stem into the water in the cup.
   c. Draw water into the pipette by releasing pressure on the bulb of the pipette.
   d. Hold the pipette by the bulb with the stem pointed up. Squeeze the bulb slightly to eliminate any air left in the bulb or stem. MAINTAIN PRESSURE ON THE BULB OF THE PIPETTE.

   e. Immediately insert the tip of the pipette's stem into the cup of water as shown in Figure.1 Release the pressure on the bulb of the pipette. The pipette will completely fill with water.

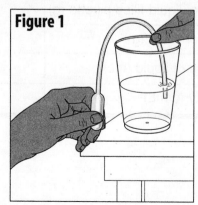

Figure 1

Laboratory Activity 1 (continued)

3. Measure the mass of the water-filled pipette. Record this value in the Data and Observations section.

Part B—Determining the Density of a Liquid

1. Completely fill the bulb of another pipette with ethanol as in Step 2 in Part A. Measure the mass of the ethanol-filled pipette. Record this value in Table 1.

Analysis

1. Calculate the mass of water in the water-filled pipette by subtracting the mass of the empty pipette from the mass of the water-filled pipette. Enter this value in the Data and Observations section.

2. The capacity of the pipette, which is the volume of the fluid that fills the pipette, can be calculated using the density of water. Because the density of water is 1.00 g/mL, a mass of 1 g of water has a volume of 1 mL. Thus, the mass of the water in the pipette is numerically equal to the capacity of the pipette. Enter the capacity of the pipette in the Data and Observations section. Record this value in Table 1 as the volume of liquid for each of the liquids used in Part B.

3. Determine the mass of each liquid by subtracting the mass of the empty pipette from the mass of the liquid-filled pipette. Record the values in Table 1.

4. Using the volumes and the masses of the liquids, calculate their densities and record them in the data table.

Data and Observations

Part A—Determining the Capacity of a Pipette

Mass of empty pipette: _____ g

Mass of water-filled pipette: _____ g

Mass of water: _____ g

Capacity of pipette: _____ mL

Part B—Determining the Density of a Liquid

Table 1

Measurement	Liquid		
	Ethanol	**Corn oil**	**Corn syrup**
1. Mass of liquid-filled pipette (g)			
2. Mass of liquid (g)			
3. Volume of liquid (mL)			
4. Density (g/mL)			

Laboratory Activity 1 (continued)

Questions and Conclusions

1. Rank the liquids by their densities starting with the least dense.

   _____

   _____

2. How does the density of water compare to the densities of the other liquids?

   _____

   _____

3. What would you observe if you poured corn oil into a beaker of water? Why?

   _____

   _____

   _____

4. The specific gravity of a substance is the ratio of the density of that substance to the density of a standard, which is water. Specific gravity is a measure of the relative density of a substance. Determine the specific gravity of ethanol, corn oil, and corn syrup.

   _____

   _____

5. Why doesn't specific gravity have units? Determine the specific gravity of ethanol, corn oil, and corn syrup.

   _____

   _____

   _____

Strategy Check

_____ Can you determine the capacity of a pipette?

_____ Can you measure the masses of several liquids?

_____ Can you calculate the densities of the liquids?

_____ Can you compare the densities of the liquids with that of water?

LAB 2 Laboratory Activity

The Behavior of Gases

Chapter 14

Because most gases are colorless, odorless, and tasteless, we tend to forget that gases are matter. Because the molecules of a gas are far apart and free to move, a gas fills its container. The volume of a gas changes with changes in its temperature and pressure. Gases expand and contract as the pressure on them changes. Gases expand when the pressure on them decreases. They contract when the pressure on them increases. The volume and pressure of a gas are inversely related. Gases also expand and contract as their temperature changes. The expansion of a gas varies directly with its temperature.

Strategy

You will observe how the volume of a gas is affected by a change in pressure.
You will observe how the volume of a gas is affected by a change in temperature.

Materials

methylene blue solution
3 small plastic cups
2 plastic microtip pipettes
water
hot plate, laboratory burner,
    or immersion heater

pliers
5 identical books
metric ruler
24-well microplate

iron or lead washer
masking tape
250-mL beaker

Procedure

Part A—Volume and Pressure of a Gas

1. Place two drops of methylene blue solution in a small plastic cup. Pour water into the cup until it is half full.
2. Fill only the bulb of the plastic pipette with this solution.
3. Seal the tip of the pipette in the following manner: Soften the tip of the pipette by holding the tip near the surface of the hot plate or near the flame of the burner. **WARNING:** *Do not place the tip of the stem on the hot plate or in the flame of the burner. Avoid coming in contact with the hot plate or the flame of the burner.* Away from the heat, squeeze the softened tip of the pipette with the pliers to seal the end. See Figure 1.

4. Place one of the books on the bulb of the pipette and measure in mm the length of the column of air trapped in the stem of the pipette. Record this value in Table 1.
5. Predict what will happen to the length of the trapped air column if another book is placed on top of the first book. Record your prediction in the Data and Observations section.
6. Place another book on top of the first book. Measure, in mm, the length of the column of trapped air and record the measure in Table 1.
7. Continue adding books one at a time, until five books are stacked on top of the pipette. After adding each book, measure the length of the column of trapped air and record the measurement in Table 1.

Figure 1

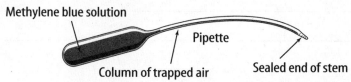

Methylene blue solution
Pipette
Column of trapped air
Sealed end of stem

Laboratory Activity 2 (continued)

Part B—Volume and Temperature of a Gas

1. Fill a well of the microplate with water. Add a few drops of methylene blue solution to the well.
2. Place an iron or lead washer over the end of the stem of the second pipette. Place the bulb in a plastic cup two-thirds filled with water at room temperature. See Figure 2.

Figure 2

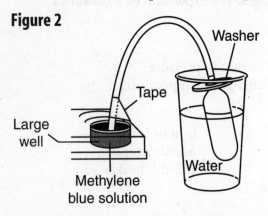

Large well

Tape

Washer

Water

Methylene blue solution

3. Bend the stem of the pipette into the solution in the well of the microplate. With the tip of the stem below the surface of the solution, tape the stem to the side of the microplate. The tip of the stem must remain below the surface of the solution during the remainder of the experiment. See Figure 2.
4. Predict what you will observe if the bulb of the pipette is gently heated. Write your prediction in the Data and Observations section.
5. Heat some water in the 250-mL beaker to a temperature of 30° C.
6. Pour the warmed water into another plastic cup until it is two-thirds full.

7. Remove the bulb of the pipette from the room-temperature water and place it in the warm water in the second cup. Immediately begin counting the bubbles that rise from the tip of the stem submerged in the well of the microplate until it stops bubbling. Record the number of bubbles and the temperature of the water in Table 2.
8. Empty the water from the first plastic cup.
9. Add some water to the beaker and heat the water to a temperature of 35° C. Pour this water into the first plastic cup until it is two-thirds full.
10. Remove the bulb of the pipette from the second cup and place it in the water in the first cup. Count the number of bubbles that rise in the well of the microplate. Record this number and the temperature of the water in Table 2. Empty the water from the second plastic cup.
11. Repeat steps 8–10 for the water that has been heated to 40°C, 45°C, and 50°C.

Analysis

1. Make a graph of your data from Part A using Graph 1. Plot the pressure on the x-axis and the length of the trapped air column on the y-axis. Label the x-axis *Pressure (books)* and the y-axis *Length (mm)*.
2. Complete the third column of Table 2. Make a graph of your data from Part B using Graph 2. Plot the temperature on the x-axis and the total number of bubbles on the y-axis. Label the x-axis *Temperature (°C)* and the y-axis *Total number of bubbles*.

Data and Observations

Part A—Volume and Pressure of a Gas

1. Prediction of length of trapped air column if the pressure on the pipette bulb is increased:

_____

_____

Laboratory Activity 2 (continued)

Table 1

Pressure (number of books)	Length of column of trapped air (mm)
1	
2	
3	
4	
5	

Part B—Volume and Temperature of a Gas

2. Prediction of observations if the air in the bubble is heated:

_____

_____

Graph 1

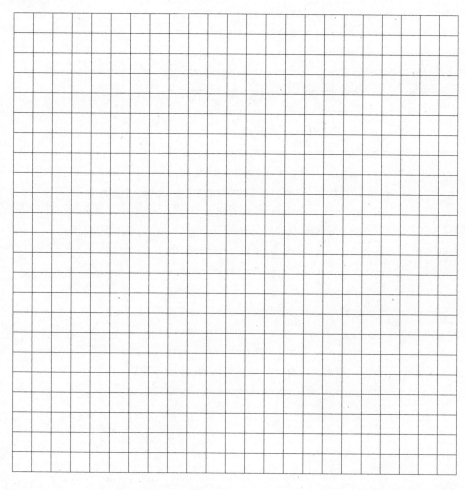

Laboratory Activity 2 (continued)

Table 2

Temperature (°C)	Number of bubbles	Total number of bubbles
_____ (room temp)		
30		
35		
40		
45		
50		

Graph 2

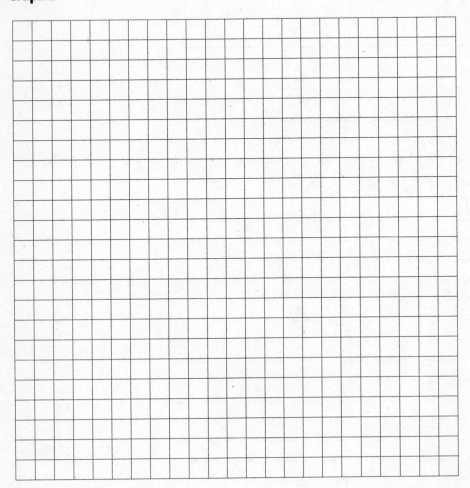

Laboratory Activity 2 (continued)

Questions and Conclusions

1. Explain how the change in the length of the column of trapped air in Part A is a measure of the change in the volume of the air trapped in the pipette.

_____

_____

_____

_____

2. Why did you have to stack identical books on the bulb of the pipette?

_____

_____

_____

_____

_____

3. What is the relationship between the volume and pressure of a gas?

_____

_____

_____

_____

4. Explain why the total number of bubbles produced is a measure of the change in volume of the air that was heated in the bulb of the pipette.

_____

_____

_____

_____

_____

Laboratory Activity 2 (continued)

5. Use your graph to predict the total number of bubbles released if the bulb of the pipette were placed in water at a temperature of 60°C.

_____

_____

6. During each 5°C temperature change, the number of bubbles released was the same. What does this indicate?

_____

_____

_____

7. What is the relationship between the volume and temperature of a gas?

_____

_____

_____

Strategy Check

_____ Can you observe how the volume of a gas is affected by a change in pressure?

_____ Can you observe how the volume of a gas is affected by a change in temperature?

Chromatography

Chromatography is a useful method for separating substances in a mixture. As you recall, the substances in a mixture are not chemically combined. Therefore, they can be separated. Chromatography can be used to separate the substances in certain mixtures because these substances dissolve at different rates.

Many mixtures, such as inks and food colorings, consist of two or more dyes. To separate the dyes, a small portion of the mixture is put on an absorbent material, such as filter paper. A liquid called a solvent is absorbed onto one end of the filter paper.

The solvent soaks the filter paper, dissolving the ink. If a dye in the ink dissolves well, it will move along the paper at the same rate as the solvent. If another dye in the ink doesn't dissolve as well, it will not move as far.

In a short time, a pattern of colors will appear on the filter paper. Each color will be a single dye that was in the ink. The distance that a component dye travels on the filter paper is a property of that dye. You can use this property to identify dyes that are found in inks of other colors.

Strategy

You will use chromatography to separate the substances in a mixture.
You will show differences in the physical properties of the substances that make up a mixture.

Materials

24-well microplate
filter paper
scissors
pencil
metric ruler
red, green, and black ink marking pens

plastic microtip pipette
ethanol
distilled water
masking tape
resealable plastic bag
paper towel

Procedure

1. Place the 24-well microplate on a flat surface. Arrange the plate so that the numbered columns are at the top and the lettered rows are at the left.
2. Cut three strips of filter paper so that each is approximately as long as the microplate and 1.5 cm wide.

3. Use a pencil to draw a line 1 cm from one end across each strip of filter paper.
4. Make a spot, using the red ink marking pen, in the middle of the pencil line on one of the strips of filter paper. After the ink has dried, apply more ink to the same spot. Allow the ink to dry. See Figure 1.

Figure 1

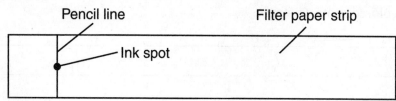

Pencil line   Filter paper strip

Ink spot

Laboratory Activity 1 (continued)

5. Repeat step 4 for the two remaining strips of filter paper. Use the green ink marker to spot one strip and the black ink marker to spot the other.

6. Half-fill the microtip pipette with ethanol. Empty the pipette into well B1 of the microplate.

7. Repeat step 6 using distilled water. Thoroughly mix the ethanol and water in the well.

8. Repeat steps 6 and 7 using wells C1 and D1.

9. Place the end of the first strip of filter paper into well B1 so that the pencil line is about 0.5 cm from the edge of the well. Do not allow the pencil line or spot to come into contact with the solution in the well. The end of the filter paper, however, must be in contact with the solution in the well.

10. Stretch the strip along the top of the microplate. Attach the end of the strip to the microplate with a small piece of tape.

11. Repeat steps 9 and 10 for the two remaining strips using wells C1 and D1.

12. Carefully place the microplate inside the plastic bag and seal the bag. See Figure 2.

13. Observe the spots on the strips of the filter paper. Record your observations in the Data and Observations section.

14. When the solvent reaches the ends of the strips, remove the plate from the plastic bag.

15. Remove the strips from the wells and allow the strips to dry on a paper towel. **WARNING:** *The dyes on the strips can easily stain your hands and clothing; do not touch the colored areas of the strips.*

16. Note the colors of the dyes on each strip. Record these colors in Table 1 for each color of ink used.

17. Attach the dried strips below the Data and Observations section.

Figure 2

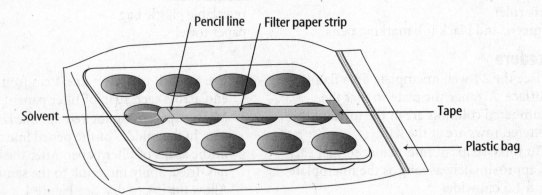

Pencil line   Filter paper strip

Solvent

Tape

Plastic bag

Data and Observations

Observations of colored spots on strips:

_____

_____

_____

_____

Laboratory Activity 1 (continued)

Table 1

Ink	Color of component dyes
Red	
Green	
Black	

Attach the dried strips of paper here.

Laboratory Activity 1 (continued)

Questions and Conclusions

1. The term *chromatography* is related to the Greek roots *chroma*, meaning color, and *graphos*, meaning written. Use the observations you made during this lab to explain how chromatography reflects the meaning of its roots.

_____

_____

_____

2. Explain if a physical or chemical change took place during the chromatography experiment.

_____

_____

3. What observations would indicate that an ink is made of a single dye?

_____

_____

4. Which component dye traveled the greatest distance for each ink?

Red ink: _____

Green ink: _____

Black ink: _____

5. A student cut out the two colored spots that she observed on the strip of filter paper that had the green ink spot. She placed the two cut-out spots into two wells of the microplate. She then added an equal amount of ethanol and distilled water to each well. She noticed that the solutions in the wells became colored. She repeated the chromatography experiment, spotting each solution on a different strip of filter paper. Predict what she will see on the strips of filter paper after the experiment. Explain your prediction.

_____

_____

_____

_____

_____

Strategy Check

_____ Can you use chromatography to separate the substances in a mixture?

Properties of Matter

Everything that has mass and takes up space is called matter. Matter exists in four different states: solid, liquid, gas, and plasma. This paper, your hand, water, and the air you breathe all consist of matter. Even the planets and stars are made of matter.

Scientists use two types of properties to describe matter. Physical properties depend on the nature of the matter. They are observed when there is no change in chemical composition. The physical properties of water describe it as a colorless, nonmagnetic liquid between the temperatures of 0°C and 100°C. Chemical properties describe the change in chemical composition of matter due to a chemical reaction. A chemical property of water is its reaction with iron to form rust.

Matter is constantly changing. A physical change involves a change in shape, temperature, state, and so on. When a material changes composition, a chemical change occurs.

Strategy

You will classify materials by states of matter.
You will identify physical and chemical properties.
You will distinguish between physical and chemical changes.

Materties ☠ 🚫 🥽 ✋

iron sample
copper sample
insulated copper wires (3)
hydrochloric acid (HCl)
toast
wood sample

magnet
lamp
test tube
chalk
dropper
rubber sample

1.5-V dry cells (2)
masking tape
test-tube rack
kitchen matches
iodine solution

WARNING: *Hydrochloric acid is corrosive, and iodine solution is poisonous. Handle these solutions with care.*

Procedure

Part A—States of Matter

1. Your teacher has set up a bottle containing different materials. Describe the state of matter for each material in the bottle. Record your observations in the Data and Observations section.

Part B—Physical Properties

1. Examine the samples of iron, wood, rubber, and copper. In Table 1, describe the physical properties listed and any other properties you can readily observe.

2. Test each sample for its attraction to a magnet. Record your observations in Table 2.

3. Use 2 fresh dry cells, 3 wires, and a small lamp to test each sample for its ability to conduct electricity. Set up the materials as shown in Figure 1. Use tape to secure each connection. Attach wires to both ends of the sample. Record the conductivity in Table 2. You will know that the sample is a conductor if the bulb lights.

Laboratory Activity 2 (continued)

Part C—Chemical Properties

1. *Safety goggles and a laboratory apron must be worn for this part of the experiment.* Add hydrochloric acid to the test tube until it is about half full. Place a small piece of chalk in the acid and observe what happens. Record your observations in Table 3.

2. Hold a burning match directly over the mouth of the test tube. Record your observations in Table 3.

3. Break a piece of toast to expose the untoasted center. Use a dropper to add a drop of iodine solution to the toasted portion of the toast. Add another drop to the untoasted center. Record your observations in Table 3.

Figure 1

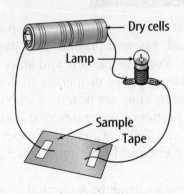

Dry cells

Lamp

Sample

Tape

Data and Observations

Part A—States of Matter

States of matter in the bottle:

_____

_____

Part B—Physical Properties

Table 1

Sample	Color	Shape	State of matter	Other properties
iron				
wood				
rubber				
copper				

Laboratory Activity 2 (continued)

Part C—Chemical Properties

Table 2

Sample	Attracted to magnet?	Conducts electricity?
iron		
wood		
rubber		
copper		

Table 3

Materials reacting	Observations
chalk and hydrochloric acid	
iodine and toasted bread	
iodine and untoasted bread	

Questions and Conclusions

1. What states of matter were visible in the bottle? What states were present but invisible in the bottle?

   _____

   _____

   _____

2. What are two physical properties that iron and copper have in common?

   _____

   _____

   _____

3. Why are your observations of the four samples descriptions of physical properties?

   _____

   _____

   _____

4. When you added chalk to hydrochloric acid, what type of change took place? How do you know?

   _____

   _____

Laboratory Activity 2 (continued)

5. List one physical property of the gas created by adding chalk to hydrochloric acid. List one chemical property of this gas.

_____

_____

_____

6. What type of change took place when iodine was dropped on the untoasted bread? How do you know?

_____

_____

_____

Strategy Check

_____ Can you classify materials by states of matter?

_____ Can you identify physical and chemical properties?

_____ Can you distinguish between physical and chemical changes?

Chemical Activity

The atoms of most chemical elements can either gain or lose electrons during reactions. Elements whose atoms lose electrons during reactions are classified as metals. Metals are found on the left side of the periodic table of elements. The tendency of an element to react chemically is called activity. The activity of a metal is a measure of how easily the metal atom loses electrons.

Strategy

You will observe chemical reactions between metals and solutions containing ions of metals.
You will compare the activities of different metals.
You will rank the metals by their activities.

Materials

96-well microplate
white paper
plastic microtip pipette
distilled water
aluminum nitrate solution, $Al(NO_3)_3aq$
copper(II) nitrate solution, $Cu(NO_3)_2aq$
iron(II) nitrate solution, $Fe(NO_3)_3aq$
magnesium nitrate solution $Mg(NO_3)_2aq$

nickel nitrate solution, $Ni(NO_3)_2aq$
zinc nitrate solution, $Zn(NO_3)_2aq$
8 1-mm × 10-mm strips of each:
    aluminum, Al; copper, Cu; iron, Fe;
    magnesium, Mg; nickel, Ni; and Zinc, Zn
paper towels
hand lens or magnifier

WARNING: *Many of these solutions are poisonous. Avoid inhaling any vapors from the solutions. These solutions can cause stains. Avoid contacting them with your skin or clothing.*

Procedure

1. Place the microplate on a piece of paper on a flat surface. Have the numbered columns of the microplate at the top and the lettered rows at the left.

2. Using the microtip pipette, place 15 drops of the aluminum nitrate solution in each of the wells A1–G1. Rinse the pipette with distilled water.

3. Place 15 drops of copper nitrate solution in each of wells A2–G2 using the pipette. Rinse the pipette with distilled water.

4. Repeat step 1 for each of the remaining solutions. Add the iron nitrate solution to wells A3–G3, the magnesium nitrate solution to wells A4–G4, the nickel nitrate solution to wells A5–G5, the zinc nitrate solution to wells A6–G6. Leave the wells in column 7 empty.

5. Carefully clean each metal strip with a paper towel.

6. Place one strip of aluminum in each of the wells A1–A7.

7. Place one strip of copper in each of wells B1–B7.

8. Repeat step 5 for the remaining metals. Add the iron strips to wells C1–C7, the magnesium strips to wells D1–D7, the nickel strips to wells E1–E7, and the zinc strips to wells F1–F7. Do not put strips in the wells in row G.

Laboratory Activity 1 (continued)

Figure 1

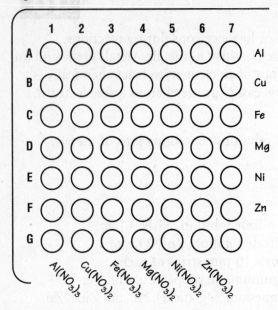

9. Figure 1 shows the metal and the solution that are in each of the wells A1–G7.
10. Wait ten minutes.
11. Use a hand lens or magnifier to observe the contents of each well. Look for a change in the color of the solution in each well by comparing it with the color of the solution in well G at the bottom of the column. Look for a change in the texture or color of the metal strip in each well by comparing it with the piece of metal in well 7 near the end of that row.

Look for the appearance of deposited materials in the bottom of the well. Each change of appearance of deposits is an indication that a chemical reaction has taken place.

12. If you see an indication of a reaction, draw a positive sign (+) in the corresponding well of the microplate shown in Figure 2 in the Data and Observations section. If you see no indication of a reaction, draw a negative sign (–) in the corresponding well of Figure 2.

Figure 2

```
     1   2   3   4   5   6   7
A   O   O   O   O   O   O   O
B   O   O   O   O   O   O   O
C   O   O   O   O   O   O   O
D   O   O   O   O   O   O   O
E   O   O   O   O   O   O   O
F   O   O   O   O   O   O   O
G   O   O   O   O   O   O   O
```

Data and Observations

Count the number of positive signs in each row of wells in Figure 2. Record the value under the corresponding metal in Table 1.

Table 1

Metal	Al	Cu	Fe	Mg	Ni	Zn
Number of reactions						

Laboratory Activity 1 (continued)

Questions and Conclusions

1. Why were solutions but no strips of metal placed in wells G1–G7?

_____

_____

_____

2. Why were strips of metal but no solutions added to wells A7–G7?

_____

_____

_____

3. Why did you clean the metal strips with the paper towel?

_____

_____

_____

_____

4. Using the number of reactions for each metal in Table 1, rank the metals from the most active to the least active.

_____

_____

_____

Strategy Check

_____ Can you determine whether or not a solution is active?

_____ Can you put metals in order based on their activities?

Modeling the Half-Life of an Isotope

Isotopes are atoms of the same element with different atomic masses. These different masses are a result of having different numbers of neutrons in their nuclei. Isotopes can be stable or unstable (radioactive). Radioactive isotopes have unstable nuclei that break down in a process called radioactive decay. During this process, the radioactive isotope is transformed into another, usually more stable, element. The amount of time it takes half the atoms of a radioactive isotope in a particular sample to change into another element is its half-life. A half-life can be a fraction of a second for one isotope or more than a billion years for another isotope, but it is always the same for any particular isotope.

Strategy

You will make a model that illustrates the half-life of an imaginary isotope.
You will graph and interpret data of the isotope's half-life.

Materials

100 pennies
plastic container with lid
timer or clock with second hand
colored pencils

Procedure

1. Place 100 pennies, each head-side up, into the container. Each penny represents an atom of an unstable isotope.
2. Place the lid securely on the container. Holding the container level, shake it vigorously for 20 seconds.
3. Set the container on the table and take off the lid. Remove only the pennies that are now in a tails-up position.
4. Count the pennies you have removed and record this number in Table 1 under *Trial 1*. Also record the number of heads-up pennies that are left.
5. Repeat steps 2 through 4 until there are no pennies left in the container.
6. Repeat steps 1 through 5 and record your data in Table 1 under *Trial 2.*
7. Calculate the averages for each time period and record these numbers in Table 1.

8. Graph the average data from Table 1 on Graph 1. Use one colored pencil to graph the number of heads-up pennies against time. Make a key for the graph that shows this color as *Radioactive Isotopes.* Using a different color of pencil, plot the number of tails-up pennies against time. In your key, show this color as *Stable Atoms.*
9. Record your averages from Table 1 again in Table 2 under *Group 1.*
10. Then, record the averages obtained by other groups in your class in Table 2.
11. Determine the totals for the combined data from all groups in Table 2.
12. Graph this combined data in Graph 2 in the same way as you graphed your group's data in step 8.

Laboratory Activity 2 (continued)

Data and Observations

Table 1

Shaking time	Trial 1 A — Number of heads-up remaining	Trial 1 B — Number of tails-up removed	Trial 2 C — Number of heads-up remaining	Trial 2 D — Number of tails-up removed	Averages — Columns A and C (H)	Averages — Columns B and D (T)
After 20 s						
After 40 s						
After 60 s						
After 80 s						
After 100 s						
After 120 s						
After 140 s						

Table 2

Group Average	Start H*	Start T*	20 s H	20 s T	40 s H	40 s T	60 s H	60 s T	80 s H	80 s T	100 s H	100 s T	120 s H	120 s T	140 s H	140 s T
Group 1	100	0														
Group 2	100	0														
Group 3	100	0														
Group 4	100	0														
Group 5	100	0														
Group 6	100	0														
Group 7	100	0														
Group 8	100	0														
Totals																

*Note: H = heads, T = tails

Laboratory Activity 2 (continued)

Graph 1

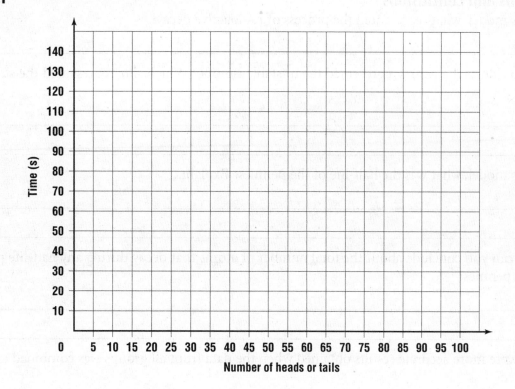

Graph 2

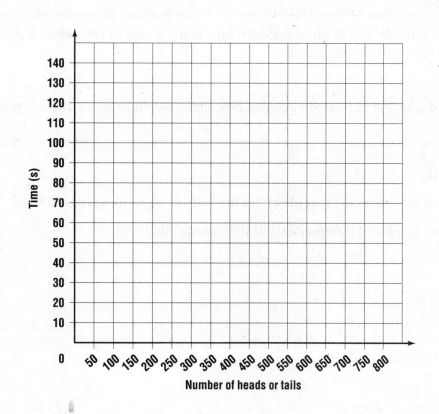

Laboratory Activity 2 (continued)

Questions and Conclusions

1. In this model, what represented the process of radioactive decay?

_____

2. Which side of the penny represented the unstable isotope? Which side represented the stable atom?

_____

_____

3. In this model, what was the half-life of the pennies? Explain.

_____

_____

4. What can you conclude about the total number of atoms that decay during any half-life period of the pennies?

_____

_____

5. Why were more accurate results obtained when the data from all groups was combined and graphed?

_____

_____

6. If your half-life model had decayed perfectly, how many atoms of the radioactive isotope should have been left after 80 seconds?

_____

7. If you started with 256 radioactive pennies, how many would be stable after 60 seconds?

_____

Strategy Check

_____ Can you make a model that illustrates the half-life of an imaginary isotope?

_____ Can you graph and interpret data of the isotope's half-life?

Preparation of Carbon Dioxide

LAB 1 Laboratory Activity

Chapter 17

When you burn a material that contains carbon, such as paper or gasoline, carbon dioxide gas is produced. You also produce carbon dioxide when your body "burns" the food you eat. You don't burn the food with a flame, however. The cells of your body combine the carbon in the food you eat with the oxygen in a reaction called oxidation. When carbon compounds are oxidized, carbon dioxide gas is produced.

Carbon dioxide gas is colorless, odorless, and tasteless. It is necessary for photosynthesis, the process by which green plants produce oxygen and glucose.

Strategy

You will observe a reaction that produces carbon dioxide gas.
You will describe the reaction that produces carbon dioxide gas.
You will observe the chemical properties of carbon dioxide gas.

Materials

metric ruler
distilled water
lime water
toothpicks
matches

hydrochloric acid solution
24-well microplate
scissors
long stem plastic pipette

forceps
marble chips
transparent tape
plastic microtip pipettes (4)

WARNING: *Hydrochloric acid is corrosive. Avoid its contact with your skin or clothing. Rinse spills with water.*

Procedure

Part A—Preparing Carbon Dioxide Gas

1. Place the microplate on a flat surface. Have the numbered columns of the microplate at the top and the lettered rows at the left.
2. Use the scissors to trim the stem of the long stem pipette to a length of 2.5 cm.
3. Using the scissors, cut a small slit in the pipette as shown in Figure 1.

Figure 1

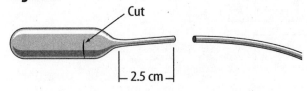

Cut

2.5 cm

4. Use the forceps to insert a small marble chip through the slit into the bulb of the pipette. Cover the slit with transparent tape to seal the bulb.Place the bulb of the pipette in well A1.
5. Make collector pipettes by cutting the stems of 2 of the microtip pipettes to lengths of 1 cm, as shown in Figure 2.

Figure 2

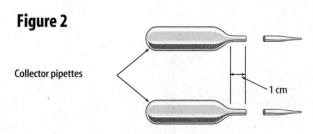

Collector pipettes

1 cm

6. Completely fill the two collector pipettes with water by holding each pipette under running water with its stem upward. Squeeze the bulb repeatedly until there is no more air in the pipette.
7. Stand the collector pipettes with their stems upward in wells C1 and C2.
8. Using an uncut microtip pipette, add about half a pipettefull of hydrochloric acid to well C3. Rinse the pipette with distilled water.
9. Take the pipette containing the marble chip from well A1 and invert it.

Laboratory Activity 1 (continued)

10. Squeeze out the air. Place the stem of the pipette in well C3 and draw the hydrochloric acid into the bulb of the pipette. Immediately invert the pipette.

11. Take the collector pipette from well C1 and invert it over the stem of the pipette containing the hydrochloric acid and marble chip. Insert the stem of the lower pipette into the stem of the collector pipette. Place the stem of the lower pipette into the bulb of the collector pipette as far as it will go. Place the pipettes into well C4 as shown in Figure 3. Allow the displaced water from the upper pipette to collect in the well.

12. Observe the reaction of the marble chip and hydrochloric acid. Record your observations in the Data and Observations section.

13. Allow the bulb and about 0.5 cm of the stem of the collector pipette to fill with gas. Remove the collector pipette and invert it. Allow the water to form a "plug" sealing the gas in the pipette as shown in Figure 4.

14. Return the collector pipette to well C1.

15. Remove the second collector pipette from well C2 and the pipette containing the hydrochloric acid and the marble chip from well C4.

16. Repeat steps 11 and 13. Return the second collector pipette to well C2.

Figure 3

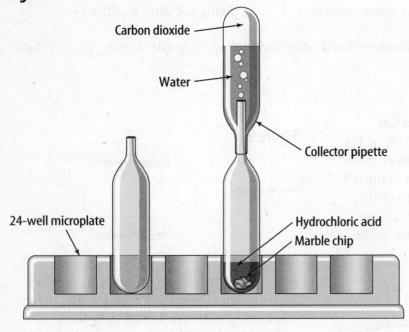

Carbon dioxide

Water

Collector pipette

24-well microplate

Hydrochloric acid

Marble chip

Figure 4

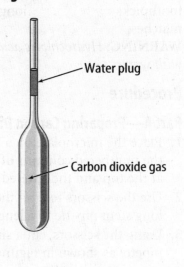

Water plug

Carbon dioxide gas

Laboratory Activity 1 (continued)

Part B—Properties of Carbon Dioxide Gas

1. Fill a clean microtip pipette with the lime water.

2. Observe the color of the lime water. Record your observations in the Data and Observations section.

3. Push the tip of the microtip pipette into the stem of the collector pipette in well C1. Push the tip through the water plug and into the bulb of the collector pipette.

4. Add about one-fourth a pipettefull of the lime water to the collector pipette. Remove the upper pipette.

5. Remove the collector pipette from well C1. Cover the tip of the pipette with your finger and shake the pipette vigorously for about 20 seconds.

6. Return the pipette to well C1. Observe the color of the solution. Record your observations in the Data and Observations section.

7. Ignite the tip of a toothpick with a match. **CAUTION:** *Use care with open flames.* Extinguish the flame, allowing the tip of the toothpick to glow.

8. Remove the water plug from the collector pipette in well C2 by gently squeezing the bulb of the pipette.

9. Immediately insert the glowing tip of the toothpick into the bulb of the collector pipette.

10. Observe the tip of the toothpick. Record your observations in the Data and Observations section.

Data and Observations

Part A—Preparing Carbon Dioxide Gas

Step 12. Observations of reaction of marble chip and hydrochloric acid

_____

_____

Part B—Properties of Carbon Dioxide Gas

Step 2. Observations of lime water

_____

_____

Step 6. Observations of solution

_____

_____

Step 10. Observations of glowing tip of toothpick in carbon dioxide gas

_____

_____

Laboratory Activity 1 (continued)

Questions and Conclusions

1. When carbon dioxide gas and lime water are mixed, calcium carbonate is formed. Describe how your observations of the reaction of lime water and carbon dioxide gas can be used to identify carbon dioxide gas.

   _____

   _____

2. Carbon dioxide gas does not support combustion. Describe how your observations of the glowing toothpick can be used to identify carbon dioxide gas.

   _____

   _____

3. When a can of a soft drink is opened, bubbles of carbon dioxide gas form. When hydrochloric acid and marble chips are mixed, bubbles of carbon dioxide gas are produced. How do the two situations differ?

   _____

   _____

   _____

   _____

Strategy Check

_____ Can you describe the reaction that produces carbon dioxide gas?

Preparation of Oxygen

Chapter 17

LAB 2 Laboratory Activity

About 20 percent of Earth's atmosphere is oxygen. Oxygen gas is colorless, odorless, and tasteless. You, as well as most living organisms, require oxygen for respiration.

On Earth, most metallic elements are found as oxides. An oxide is a compound containing oxygen and another element. One oxide with which you are familiar is silicon dioxide—sand. Sand and water are the most common compounds of oxygen on this planet's surface.

Strategy

You will observe a reaction that produces oxygen gas.
You will describe the reaction that produces oxygen gas.
You will observe the chemical properties of oxygen gas.

Materials

24-well microplate	cobalt nitrate solution	toothpicks
plastic microtip pipettes (3)	metric ruler	matches
household bleach solution	scissors	
distilled water	long-stem plastic pipette	

WARNING: *Bleach and cobalt nitrate solution can cause stains; avoid contact with your skin or clothing. Rinse spills with water.*

Procedure

Part A—Preparing Oxygen Gas

1. Place the microplate on a flat surface. Have the numbered columns of the microplate at the top and the lettered rows at the left.

2. Using a clean microtip pipette, add 30 drops of the household bleach to well A1. Rinse the pipette with distilled water.

3. Using the microtip pipette, add 10 drops of the cobalt nitrate solution to well A2. Rinse the pipette with distilled water.

4. Make collector pipettes by cutting the stems of two of the microtip pipettes to lengths of 1 cm, as shown in Figure 1.

5. Completely fill the two collector pipettes with water by holding each pipette under running water with its stem upward. Squeeze the bulb repeatedly until there is no more air in the pipette.

6. Stand the collector pipettes with their stems upward in wells C1 and C2.

Figure 1

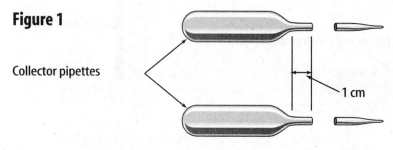

Collector pipettes

1 cm

Figure 2

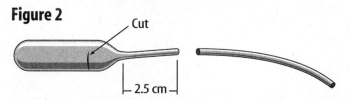

Cut

2.5 cm

Laboratory Activity 2 (continued)

7. Use the scissors to trim the stem of the long stem pipette to a length of 2.5 cm as shown in Figure 2.

8. Using this pipette, draw up all the bleach solution from well A1 into the bulb of the pipette.

9. Hold the pipette with the stem upward. Gently squeeze out the air. While still squeezing the bulb, invert the pipette and place the stem into well A2. Draw the cobalt nitrate solution into the bulb of the pipette. Immediately invert the pipette.

10. Take the collector pipette from well C1 and invert it over the stem of the pipette containing the bleach and cobalt nitrate solutions. Insert the stem of the lower pipette into the stem of the collector pipette. Place the stem of the lower pipette into the bulb of the collector pipette as far as it will go. Place the pipettes into well C4 as shown in Figure 3. Allow the displaced water from the upper pipette to collect in the well.

11. Observe the reaction of the bleach and cobalt nitrate solutions. Record your observations in the Data and Observations section.

12. Allow the bulb and about 0.5 cm of the stem of the collector pipette to fill with gas. Remove the collector pipette and invert it. Allow the water to form a "plug" sealing the gas in the pipette as shown in Figure 4.

13. Return the collector pipette to well C1.

14. Remove the second collector pipette from well C2 and the pipette containing the bleach and cobalt nitrate solutions from well C4.

15. Repeat steps 10 and 12. Return the collector pipette to well C2.

Figure 3

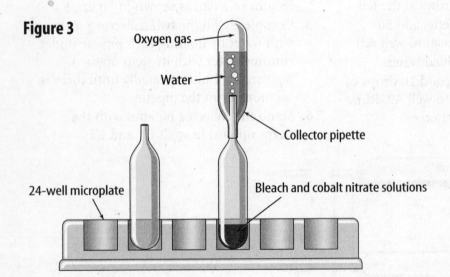

Oxygen gas

Water

Collector pipette

24-well microplate

Bleach and cobalt nitrate solutions

Figure 4

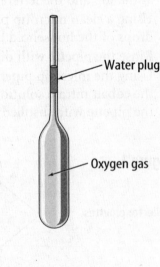

Water plug

Oxygen gas

Laboratory Activity 2 (continued)

Part B—Properties of Oxygen Gas

1. Ignite the tip of a toothpick with a match. **CAUTION:** *Use care with open flames.* Extinguish the flame.
2. Remove the water plug from the collector pipette in well C1 by gently squeezing the bulb of the pipette.
3. Immediately insert the glowing tip of the toothpick into the bulb of the collector pipette.

4. Observe the reaction. Record your observations in the Data and Observations section.
5. Repeat steps 1 through 4 for the second collector pipette in well C2.

Data and Observations

Part A—Preparing Oxygen Gas

Step 11. Observations of reaction of bleach and cobalt nitrate solutions:

_____

_____

Part B—Properties of Oxygen Gas

Step 4. Observations of glowing toothpick in presence of oxygen gas:

_____

_____

Questions and Conclusions

1. Describe how your observations of the reaction of the glowing toothpick and oxygen gas demonstrate a property of oxygen gas.

_____

_____

_____

2. What is the chemical formula of oxygen gas?

_____

3. The wood of the toothpick contains carbon compounds. What substances are formed when these carbon compounds burn?

_____

Laboratory Activity 2 (continued)

4. You observed the chemical reaction of sodium hypochlorite which is found in bleach, and cobalt nitrate solutions. The chemical formula for sodium hypochlorite is NaOCl. The chemical formula for cobalt nitrate is $Co(NO_3)_2$.

   a. What elements are in each compound?

   _____

   _____

   b. How many oxygen atoms are in each compound?

   _____

Strategy Check

_____ Can you describe the reaction that produces oxygen gas?

The Five Solutions Problem

Do you recall the seven dwarfs in the story of Snow White? Their names reflected their behaviors. One could recognize them by their actions. Substances can also be identified by their behaviors. One way of identifying a substance is by observing how it reacts with other known substances. In any chemical reaction, new substances are produced. State, color, and other physical properties of substances produced in a chemical reaction can help identify the substances that reacted.

In this experiment, you will classify solutions by how the substances in the solutions react. Using this classification, you will identify an unlabeled sample of one of these solutions.

Strategy

You will observe the reactions of five different known solutions, two at a time.
You will classify your observations.
You will identify an unlabeled sample of one of these solutions.

Materials

96-well microplate
white paper
7 plastic microtip pipettes
dilute hydrochloric acid, HCl(aq)
iron(III) nitrate solution, Fe(NO$_3$)$_3$(aq)
silver nitrate solution, AgNO$_3$(aq)

sodium carbonate solution, Na$_2$CO$_3$(aq)
sodium iodide solution, NaI(aq)
plastic cup
distilled water
sample of unknown solution 1, 2, 3, 4, or 5

WARNING: *Many of these solutions are poisonous. Avoid inhaling any vapors from the solutions. Silver nitrate solution and sodium iodide solution can cause stains. Avoid any contact between them and your skin or clothing.*

Procedure

Part A—Observing Reactions of Known Solutions

1. Wear aprons, gloves, and goggles during this experiment.
2. Place the microplate on a piece of white paper on a flat surface. Have the numbered columns of the microplate at the top and the lettered rows at the left.
3. Using a microtip pipette, place four drops of the hydrochloric acid solution in each of wells A1 through F1.
4. Using a clean pipette, place four drops of the iron(III) nitrate solution in each of wells A2 through F2.
5. Repeat step 4 for each of the remaining four solutions. Use a clean pipette for each.

Place the silver nitrate solution in wells A3 through F3, the sodium carbonate solution in wells A4 through F4, and the sodium iodide solution in wells A5 through F5.

6. Fill the plastic cup with distilled water, and thoroughly rinse each pipette. Discard the water.
7. Add four drops of the hydrochloric acid solution to each of wells A1 through A5.
8. Using another clean pipette, add four drops of iron(III) nitrate to each of wells B1 through B5.

Laboratory Activity 1 (continued)

9. Repeat step 8 for the remaining solutions. Use a clean pipette for each solution. Add the silver nitrate solution to wells C1 through C5, the sodium carbonate solution to wells D1 through D5, and the sodium iodide solution to wells E1 through E5. Figure 1 shows the solutions in each of the wells A1 through E5.

10. Observe the contents of each well. Note any changes in the physical properties of the substances in each well. Record your observations in Table 1.

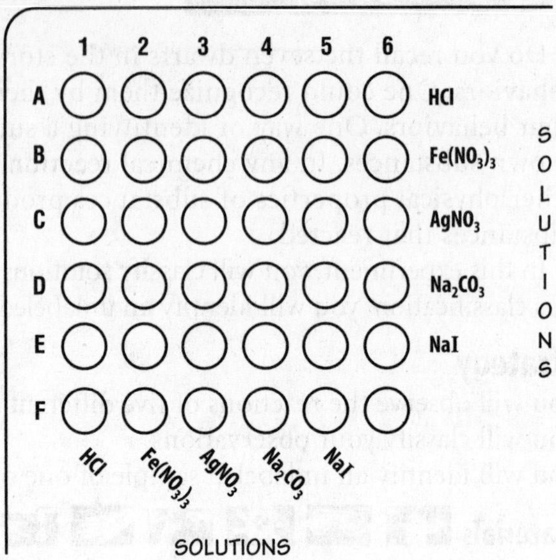

Part B—Identifying an Unknown Solution

1. Obtain a small sample of an unknown solution from your teacher. Record the number of the solution sample in the first column of Table 2.

2. Use a clean microtip pipette to add four drops of the sample solution to each of the wells F1 through F5.

3. Observe the contents of each well. Note any changes in the physical properties of the contents in each well. Record your observations in Table 2.

4. Compare the changes that occurred in wells containing the unknown solution with the changes that occurred in wells containing the known solutions.

Figure 1

SOLUTIONS

Data and Observations

Part A—Observing Reactions of Known Solutions

Table 1

Solution in microplate	Solution added				
	HCl	Fe(NO₃)₃	AgNO₃	Na₂CO₃	NaI
1. HCl					
2. Fe(NO₃)₃					
3. AgNO₃					
4. Na₂CO₃					
5. NaI					

Laboratory Activity 1 (continued)

Part B—Identifying an Unknown Solution

Table 2

Unknown solution	Solution				
	HCl	Fe(NO$_3$)$_3$	AgNO$_3$	Na$_2$CO$_3$	NaI

Questions and Conclusions

1. What is the identity of the sample solution?

_____

_____

_____

_____

_____

2. What properties of the substances that were formed helped you to identify your sample solution?

_____

_____

_____

_____

_____

3. How did the reactions between the solutions in wells A1 through E5 help you to identify the sample solution?

_____

_____

_____

_____

_____

Laboratory Activity 1 (continued)

4. Could you use the results of your observations in Part A to identify a solution that is not one of the five solutions? Explain.

_____

_____

_____

_____

_____

Strategy Check

_____ Can you classify your observations?

_____ Can you identify an unlabeled sample by the way it reacts with known substances?

Laboratory Activity 2

Investigating Covalent and Ionic Bonds

All substances are made of atoms. Some of the physical and chemical properties of a substance are determined by the chemical bonds that hold its atoms together. In this experiment you will investigate the properties of compounds formed by two types of chemical bonds—covalent bonds and ionic bonds.

The atoms of covalent compounds are held together by covalent bonds. A covalent bond forms when two atoms share electrons. In other substances, atoms transfer electrons and form ions. An ion is an atom that has gained or lost electrons. In ionic compounds, the ions are held together by ionic bonds.

Solutions of ionic compounds can conduct an electric current. These solutions of covalent compounds conduct an electric current. A measure of how well a solution can carry an electric current is called conductivity.

Strategy

You will determine the conductivity of several solutions.
You will classify the compounds that were dissolved in the solutions as ionic compounds or covalent compounds.

Materials

9-V battery and battery clip
10-cm × 10-cm cardboard sheet
masking tape
4 alligator clips
1000-Ω resistor
LED (light-emitting diode)
2 20-cm lengths of insulated copper wire
24-well microplate

7 plastic pipettes
sulfuric acid solution, $H_2SO_4(aq)$
sodium chloride solution, $NaCl(aq)$
sodium hydroxide solution, $NaOH(aq)$
silver nitrate solution, $AgNO_3(aq)$
glucose solution, $C_6H_{12}O_6(aq)$
glycerol solution, $C_3H_8O_3(aq)$
distilled water

WARNING: *Sulfuric acid and sodium hydroxide can cause burns. Silver nitrate can cause stains. Avoid inhaling any vapors from the solutions. Avoid any contact between the solutions and your skin or clothing.*

Procedure

Part A—Constructing a Conductivity Tester

1. Attach the 9-V battery clip to the 9-V battery. Use tape to attach the battery securely to the cardboard sheet, as shown in Figure 1.

2. Attach an alligator clip to one of the lead wires of the 1000-Ω resistor. Connect the same alligator clip to the red lead wire of the battery clip. Tape the resistor and alligator clip to the cardboard sheet as shown in Figure 2.

Figure 1

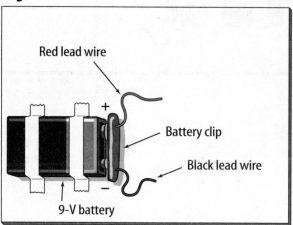

Red lead wire
Battery clip
Black lead wire
9-V battery

Laboratory Activity 2 (continued)

3. Attach an alligator clip to the *long* lead wire of the light-emitting diode (LED). Connect this alligator clip to the second wire of the 1000-Ω resistor. Tape the alligator clip to the cardboard sheet.

4. Attach an alligator clip to the *short* lead wire of the LED. Connect this alligator clip to one end of the insulated copper wires. Tape the alligator clip to the cardboard sheet as shown in Figure 3.

5. Attach the last alligator clip to one end of the second insulated copper wire. Connect the alligator clip to the *black* lead wire of the battery clip. Tape the alligator clip to the cardboard sheet as shown in Figure 4.

6. Check to be certain that the alligator clips, resistor, and battery are securely taped to the cardboard sheet and that the clips are not touching one another.

7. Have your teacher check your conductivity tester.

8. Touch the two ends of the two insulated copper wires, and observe that the LED glows.

Figure 2

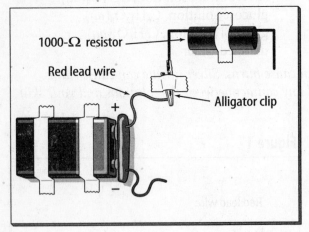

1000-Ω resistor

Red lead wire

Alligator clip

Figure 3

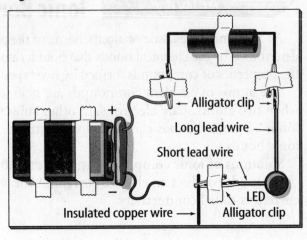

Alligator clip

Long lead wire

Short lead wire

LED

Insulated copper wire

Alligator clip

Figure 4

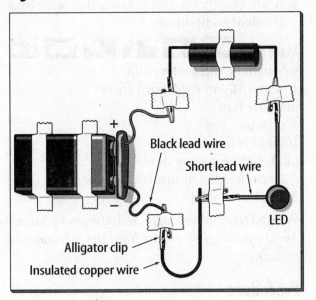

Black lead wire

Short lead wire

LED

Alligator clip

Insulated copper wire

Laboratory Activity 2 (continued)

Part B—Testing the Conductivity of a Solution

1. Wear an apron, gloves, and goggles for Part B of the experiment.
2. Place the microplate on a flat surface. Have the numbered columns of the microplate at the top and the lettered rows at the left.
3. Using a clean pipette, add a pipettefull of the sulfuric acid solution to well A1.
4. Using another clean pipette, add a pipettefull of the sodium chloride solution to well A2.
5. Repeat step 4 for each remaining solution. Use a clean pipette for each solution. Add the sodium hydroxide solution to well A3, the silver nitrate solution to well A4, the glucose solution to well A5, and the glycerol solution to well A6.
6. Using a clean pipette, add a pipettefull of distilled water to well A7. Figure 5 shows the contents of each of the wells A1 through A7.
7. Place the exposed ends of the two insulated copper wires into the solution in well A1, positioning the wires so they are at opposite sides of the well. Be sure that the exposed ends of the wires are completely submerged.
8. Observe the LED. Use the brightness of the LED as an indication of the conductivity of the solution. Rate the conductivity of the solution using the following symbols: + (good conductivity); – (fair conductivity); or 0 (no conductivity). Record your rating in the corresponding well of the microplate shown in Figure 6.
9. Remove the wires and dry the ends of the wires with a paper towel.
10. Repeat steps 6 through 9 for each remaining solution and the distilled water.

Figure 5

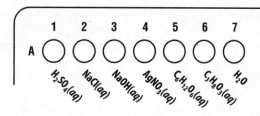

Data and Observations

Figure 6

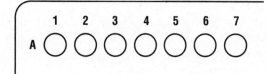

Laboratory Activity 2 (continued)

Questions and Conclusions

1. What is the conductivity of distilled water?

_____

2. Why was the conductivity of the distilled water measured?

_____

_____

3. What characteristic is common to the compounds that produce solutions that can conduct electricity?

_____

_____

4. What characteristic is shared by the compounds that produce solutions that do not conduct an electric current?

_____

_____

5. How do the conductivities of solutions of ionic compounds and covalent compounds compare?

_____

_____

_____

Strategy Check

_____ Can you determine the conductivity of solutions?

_____ Can you classify compounds in solutions as ionic or covalent?

Conservation of Mass

In a chemical reaction, the total mass of the substances formed by the reaction is equal to the total mass of the substances that reacted. This principle is called the law of conservation of mass, which states that matter is not created or destroyed during a chemical reaction.

In this experiment, sodium hydrogen carbonate, $NaHCO_3$ (baking soda), will react with hydrochloric acid, HCl. The substances formed by this reaction are sodium chloride, NaCl; water, H_2O; and carbon dioxide gas, CO_2.

Strategy

You will show that new substances are formed in a chemical reaction.

You will show the conservation of mass during a chemical reaction.

Materials

sealable plastic sandwich bag containing sodium hydrogen carbonate, $NaHCO_3$
hydrochloric acid, HCl
plastic pipette
paper towel
metric balance

Procedure

1. Obtain the plastic sandwich bag containing a small amount of sodium hydrogen carbonate.

2. Fill the pipette with the hydrochloric acid solution. Use a paper towel to wipe away any acid that might be on the outside of the pipette. Discard the paper towel. **WARNING:** *Hydrochloric acid is corrosive. Handle with care.*

Figure 1

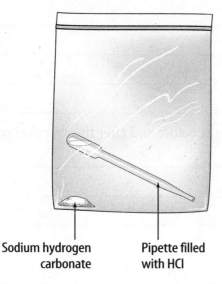

Sodium hydrogen   Pipette filled
carbonate   with HCl

3. Carefully place the pipette in the bag. Press the bag gently to eliminate as much air as possible. Be careful not to press the bulb of the pipette. Seal the bag. See Figure 1.

4. Measure the mass of the sealed plastic bag using the metric balance. Record this value in the Data and Observations section.

5. Remove the plastic bag from the balance. Without opening the bag, direct the stem of the pipette into the sodium hydrogen carbonate. Press the bulb of the pipette and allow the hydrochloric acid to react with the sodium hydrogen carbonate. Make sure that all the acid mixes with the sodium hydrogen carbonate.

6. Observe the contents of the bag for several minutes. Record your observations in the Data and Observations section.

7. After several minutes, measure the mass of the sealed plastic bag and its contents. Record this value in the Data and Observations section.

Laboratory Activity 1 (continued)

Data and Observations

Table 1

Mass of plastic bag before reaction (in grams)	
Observations from Step 6	
Mass of plastic bag after reaction (in grams)	

Questions and Conclusions

1. Why was it important for the plastic bag to be sealed?

_____

_____

_____

_____

_____

2. What did you observe that indicated that a chemical reaction took place?

_____

_____

_____

_____

3. Compare the mass of the plastic bag and its contents before and after the chemical reaction.

_____

_____

_____

_____

Laboratory Activity 1 (continued)

4. Does your comparison in Question 3 confirm the conservation of mass during this chemical reaction? Explain.

_____

_____

_____

_____

_____

Strategy Check

_____ Can you demonstrate that new substances are formed in a chemical reaction?

_____ Can you show the conservation of mass during a chemical reaction?

Chemical Reactions

The changes that occur during a chemical reaction are represented by a chemical equation. In an equation, chemical symbols represent the substances involved. The reactants are the substances that react. The products are the substances formed from the reaction. For example, reaction of the elements sodium and chlorine to produce sodium chloride is shown by the following chemical equation.

$$2Na(s) + Cl_2(g) \rightarrow 2NaCl(s)$$

reactants   *product*

In a synthesis reaction, two or more substances react to form a new substance. You may think of a synthesis reaction as putting substances together to produce a new substance. The synthesis reaction that produces hydrogen peroxide is given by the equation below.

$$2H_2O(l) + O_2(g) \rightarrow 2H_2O_2(l)$$ *Synthesis reaction*

A decomposition reaction produces several products from the breakdown of a single compound. This process is similar to breaking a single compound into several simpler compounds and/or elements.

$$2H_2O(l) \rightarrow 2H_2(g) + O_2(g)$$ *Decomposition reaction*

In a single-displacement reaction, one element replaces another element in a compound. In the following reaction carbon displaces the hydrogen in water, forming gaseous carbon monoxide, and hydrogen is released as hydrogen gas.

$$H_2O(l) + C(s) \rightarrow H_2(g) + CO(g)$$ *Single-displacement reaction*

Strategy

You will recognize the reactants and products of a chemical reaction.
You will identify the type of chemical reaction you observe.
You will write a word equation for a chemical reaction.
You will write a balanced chemical equation using chemical symbols.

Materials

Part A	Part B	Part C
aluminum foil	aluminum foil	string
burner	burner	iron nail, Fe
matches	matches	beaker
steel wool, Fe	test tube	copper (II), sulfate solution, $CuSO_4$
tongs	spoon	watch or clock
	baking soda, $NaHCO_3$	paper towel
	test tube holder	

WARNING: *Copper(II) sulfate solution is poisonous. Handle with care. Wear goggles and apron.*

Laboratory Activity 2 (continued)

Procedure

Part A—Synthesis Reaction

1. Protect the table with a sheet of aluminum foil. Place the burner in the center of the foil. Light the burner. **WARNING:** *Stay clear of the flame.*

2. Observe the color of the steel wool. Record your observations in the Data and Observations section.

3. Predict if there will be any changes in the steel wool if it is heated in the flame. Write your prediction in the Data and Observations section.

4. Hold the steel wool (containing iron, Fe) with the tongs over the flame as shown in Figure 1. As the steel wool burns, record any changes you observe.

5. Take the steel wool out of the flame and let it cool. Record your observations.

Figure 1

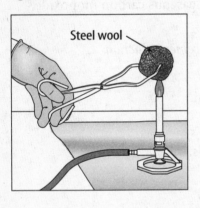

Steel wool

Part B—Decomposition Reaction

1. Set up a burner as in step 1 of Part A.

2. Place a spoonful of baking soda, $NaHCO_3$, in a test tube. In the Data and Observations section, write your prediction of what will happen as the baking soda is heated. Use the test-tube holder to heat the test tube in the flame, as shown in Figure 2. Do not point the mouth of the test tube at anyone.

Figure 2

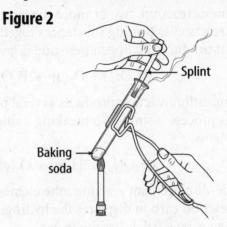

Splint

Baking soda

3. Record the description and colors of the products formed inside the tube.

4. Test for the presence of CO_2. Light a wooden splint. Hold the flaming splint in the mouth of the test tube. If the flame of the splint goes out, CO_2 is present. Record your observations of the products of this reaction.

Laboratory Activity 2 (continued)

Part C—Single Displacement Reaction

Figure 3

1. Tie a string around the nail. Fill a beaker about half full with the $CuSO_4$ solution. Record the colors of the nail and the $CuSO_4$ solution in Table 1. **WARNING:** *The $CuSO_4$ solution is toxic. Handle with care.*

2. Predict what changes will happen to the appearance of the nail and the solution when mixed. Record your prediction in the Data and Observations section. Dip the nail in the $CuSO_4$ as shown in Figure 3. After 5 minutes, pull the nail from the solution and place it on a paper towel. Record the colors of the nail and the solution in Table 1.

3. Put the nail back into the solution and observe further color changes.

Data and Observations

Part A—Synthesis Reaction

1. Prediction of changes in heated steel wool:

_____

2. Color of steel wool before burning:

_____

3. Color of burned steel wool:

_____

Part B—Decomposition Reaction

1. Prediction of changes in heated baking soda:

_____

2. Description of deposits inside heated test tube:

_____

3. Observations of flaming splint:

_____

Part C—Single Displacement Reaction

1. Prediction of changes in nail and $CuSO_4$ solution:

_____

Laboratory Activity 2 (continued)

Table 1

Observation time	Color of nail	Color of CuSO₄ solution
Before reaction		
After reaction		

Questions and Conclusions

1. Write a word equation to describe the reaction of the heated steel wool and oxygen.

   _____ plus _____, in the presence of heat,

   yields _____.

2. Write a balanced equation using chemical symbols for the synthesis reaction of iron and oxygen.

   _____

3. Write a word equation to describe the decomposition reaction of baking soda.

   _____ yields _____ plus _____ plus water.

4. Write a chemical equation using symbols for the decomposition of sodium bicarbonate, or

   baking soda. _____

5. Write a word equation to describe the single-displacement reaction of iron and copper sulfate.

   _____

6. Write a chemical equation using symbols for the single-displacement reaction of iron and

   copper(II) sulfate. _____

Strategy Check

_____ Can you recognize the reactants and products of a chemical reaction?

_____ Can you identify the type of chemical reaction you observe?

_____ Can you write a word equation for a chemical reaction?

_____ Can you write a balanced chemical equation using chemical symbols?

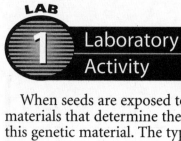

The Effect of Radiation on Seeds

Chapter 20

When seeds are exposed to nuclear radiation, changes may be observed. Seeds contain genetic materials that determine the characteristics of the plants produced from them. Radiation can alter this genetic material. The type of seeds and the amount of radiation absorbed determine the extent of this alteration.

Strategy

You will grow plants from seeds that have been exposed to different amounts of nuclear radiation.
You will observe and record the growth patterns of the plants during a period of a week.
You will use the results of your experiment to discuss some of the possible effects of exposure to nuclear radiation.

Materials

seeds that have received different amounts of radiation
seeds that have not been irradiated

potting soil
boxes or containers for planting

Procedure

1. It is important that all seeds are planted and grown under the same conditions. Plant the seeds according to your teacher's instructions. Plant one container of untreated seeds. Label this container *1*. Carefully label each of the remaining containers. In Table 1, record the number of each container and the amount of radiation the seeds planted in it received.

2. Place the containers in a location away from drafts where they can receive as much light as possible. Keep the soil moist, but not wet, at all times.

3. As soon as the first seeds sprout, start recording your observations in Table 2.

Observe the seeds at regular intervals for a week. If necessary, continue Table 2 on a separate sheet of paper. Watch for variations in sprouting and growth rates and differences in size, color, shape, number, and location of the stems and leaves. Remember, it is important to make an entry in the table for each container at every observation date, even if you report no change.

4. In the space provided in the Data and Observations section, make sketches of your plants and show any variation in growth patterns.

Data and Observations

Table 1

Container number	Amount of radiation
1	no radiation

Laboratory Activity 1 (continued)

Table 2

Date	Container number			

Plant Sketches

Laboratory Activity 1 (continued)

Questions and Conclusions

1. Why did you plant seeds that were not exposed to nuclear radiation?

_____

_____

2. What pattern or trends did you observe as the seeds sprouted?

_____

_____

_____

3. What patterns or trends did you observe in the growth rate of the plants?

_____

_____

_____

_____

4. What relationship can be seen between the amount or time of radiation exposure and the following:

maximum height of plants

_____

size of leaves

_____

color of leaves

_____

shape of leaves

_____

number of leaves

_____

placement of leaves

_____

other variations that you observed

_____

_____

Laboratory Activity 1 (continued)

5. What characteristics of the plants seem unaffected?

_____

_____

_____

6. What conclusions can you make based on the results of this experiment?

_____

_____

_____

7. What predictions can you make based on the results of this experiment?

_____

_____

_____

Strategy Check

_____ Can you grow plants from seeds that have been exposed to different amounts of nuclear radiation?

_____ Can you observe and record the growth patterns of the plants during a period of weeks?

_____ Can you use the results of your experiment to discuss some possible effects of exposure to nuclear radiation?

LAB 2 Laboratory Activity

Radioactive Decay— A Simulation

    Certain elements are made up of atoms whose nuclei are naturally unstable. The atoms of these elements are said to be radioactive. The nucleus of a radioactive atom will decay into the nucleus of another element by emitting particles of radiation. It is impossible to predict when the nucleus of an individual radioactive atom will decay. However, if a large number of nuclei are present in a sample, it is possible to predict the time period in which half the nuclei in the sample will decay. This time period is called the half-life of the element.

    Radioactive materials are harmful to living tissues. Their half-lives are difficult to measure without taking safety precautions. To eliminate these problems, you will simulate the decay of unstable nuclei by using harmless materials that are easy to observe. In this experiment you will use dried split peas to represent the unstable nuclei of one element. Dried lima beans will represent the stable nuclei of another element. Your observations will allow you to make a mental model of how the nuclei of radioactive atoms decay.

Strategy

You will simulate the decay of a radioactive element.
You will graph the results of the simulated decay.
You will determine the half-life of the element.

Materials

small bag of dried split peas
250-mL beaker
large pizza or baking tray
bag of dried lima beans

Procedure

1. Count out 200 dried split peas and place them in a beaker.

2. Record the number of split peas in Table 1 as Observation 0.

3. Place the pizza or baking tray on a flat surface.

4. Hold the beaker over the tray and sprinkle the split peas onto the tray. Try to produce a single layer of split peas on the tray.

5. Remove all the split peas that have NOT landed on the flat side down. Count the split peas that you have removed and return them to the bag. Replace the number of peas that you have removed from the tray with an equal number of lima beans. Count the number of peas and the number of lima beans on the tray. Record these values in Table 1 as Observation 1.

6. Scoop the peas and beans from the tray and place them into the beaker.

7. Predict how many split peas you will remove if you repeat steps 4 and 5. Enter your predictions in the Data and Observations section.

8. Repeat steps 4 through 6, recording your data in the data table as Observation 2.

9. Predict how many observations you will have to make until there are no split peas remaining. Enter your prediction in the Data and Observations section.

10. Repeat steps 4 through 6 until there are no split peas remaining.

Laboratory Activity 2 (continued)

Data and Observations

Table 1

Observation	Time (minutes)	Split peas	Lima beans

Prediction of number of split peas removed:

_____

Prediction of number of observations until there are no split peas remaining:

_____

Laboratory Activity 2 (continued)

Analysis

In this experiment each split pea represents the nucleus of an atom of radioactive element A. A split pea that has landed flat side down represents the nucleus of an atom of radioactive element A that has not yet decayed. Each split pea that has NOT landed flat side down represents the nucleus of element A that has decayed. Each lima bean represents the nucleus of an element B that was formed by the decay of the nucleus of an element A .

Assume that the time period between each observation was 5 minutes. Observation 1 will have been made at 5 minutes, observation 2 at 10 minutes, and so on. Complete the time column in Table 1.

1. Use Graph 1 below to graph the results of your experiment. Plot on one axis the number of the nuclei of element A atoms remaining after each observation. Plot the time of this observation on the other axis. Determine which variable should be represented by each axis.
2. Use Graph 1 to construct another graph. Plot on one axis the number of nuclei of element B atoms remaining after each observation. Plot the time of the observation on the other axis.
3. Determine the appropriate half-life of element A from your graph.

Graph 1

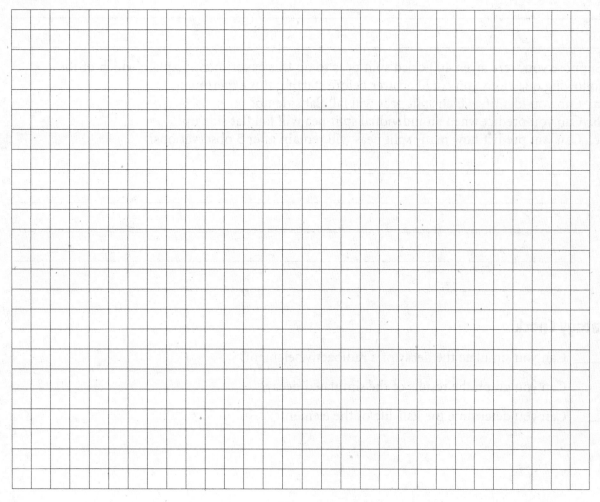

Laboratory Activity 2 (continued)

Questions and Conclusions

1. What is the approximate half-life of element A?

_____

2. Use your graph to determine the number of element A nuclei remaining after 2 half-lives, and after 3 half-lives.

_____

3. Why did you replace split peas but not lima beans during this experiment?

_____

_____

_____

4. The two graphs that you constructed look like mirror-images. Explain why this is so.

_____

_____

_____

5. Suppose you were given 400 dried split peas to do this experiment. Explain which of the following questions you could answer before starting this experiment.
   a. Can you identify which split peas will fall flat side up?
   b. Can you predict when an individual split pea will fall flat side up?
   c. Can you predict how many split peas will remain after 3 observations?

_____

_____

_____

_____

_____

Strategy Check

_____ Can you simulate the decay of a radioactive element?

_____ Can you graph the results of the simulated decay?

_____ Can you determine the half-life of the element?

Examining Properties of Solutions

To make a saltwater solution, you can use either table salt or rock salt. If the mass of each sample is the same, the salt with the greater surface area—table salt—will dissolve faster. Other factors affect the speed of the dissolving process. For example, temperature and stirring will slow down or speed up the dissolving of solute. In addition, the speed at which gases dissolve is affected by changes in pressure.

Strategy

You will explain the effects of particle size, temperature, and stirring on a solid in solution.
You will explain the effects of temperature, stirring, heating, and pressure on a gas in solution.

Materials

clear plastic cups (6)
graduated cylinder (100-mL)
table salt (3 g)
rock salt (3 g)
paper towels
stirring rod

watch with second hand or seconds mode
bottle of soda water, unopened
beaker (500-mL)
hot tap water
cold water

WARNING: *Do not taste, eat, or drink any materials used in the lab.*

Procedure

Part A—Solid in Solution

1. Label the six plastic cups A through F. Use the graduated cylinder to add 100 mL of hot tap water each to cups A and B. Add 100 mL of cold water each to cups C, D, E, and F.

2. Divide each type of salt into three equal samples.

3. Add a salt sample to each cup (one at a time) as indicated in Figure 1.

When adding each sample, observe closely and record the time required for the salt to dissolve completely. See Figure 1. When no salt particles are visible, record the time for that sample in Table 1.

4. Rate the salt samples from fastest to slowest in dissolving. Give the fastest dissolving sample a rating of 1, the slowest, a 6. Record your ratings in Table 1.

Figure 1

	A	B	C	D	E	F
Type of salt	table	rock	table	rock	table	rock
Water temperature	hot	hot	cold	cold	cold	cold
Stirred?	no	no	no	no	yes	yes

Laboratory Activity 1 (continued)

Part B—Gas in Solution

1. Rinse cups A, B, and C with water.
2. Observe the unopened bottle of soda water. Open the bottle and observe it again. Compare your observations, and record your comparison in Part B of the Data and Observations section.
3. Pour hot water from the tap into the 500-mL beaker until it is about half full.

4. Add 25 mL of soda water to each of the three cups. Stir the soda water in cup B. See Figure 2. Place cup C in the beaker of hot water. Leave cup A as your control. Compare the speed of bubbling in each cup. Record your observations in Table 2.

Figure 2

  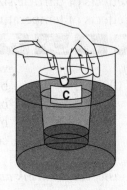

Data and Observations

Part A—Solid in Solution

Table 1

Cup	Salt sample	Water conditions	Time (s)	Rating
A	table salt	hot		
B	rock salt	hot		
C	table salt	cold		
D	rock salt	cold		
E	table salt	cold, stirred		
F	rock salt	cold, stirred		

Laboratory Activity 1 (continued)

Part B—Gas in Solution
Observations of unopened and opened bottle:

_____

_____

_____

Table 2

Cup	Soda conditions	Observations and comparison of bubbling
A	control	
B	stirred	
C	heated	

Questions and Conclusions

1. How does particle size affect the rate at which salt dissolves in water?

_____

_____

_____

_____

2. How does temperature affect the speed of the dissolving process of salt in water?

_____

_____

_____

_____

3. How does stirring affect the speed of the dissolving process of salt in water?

_____

_____

_____

_____

Laboratory Activity 1 (continued)

4. How did you create a pressure change in the bottle of soda water? What happened as a result of this pressure change?

_____

_____

_____

_____

5. What factors cause the speed of bubbling in soda water to increase?

_____

_____

_____

6. Most soft drinks contain dissolved CO_2. Sometimes when you shake a bottle of soft drink and then open it, the soft drink shoots into the air. Explain why this happens.

_____

_____

_____

_____

_____

_____

Strategy Check

_____ Can you explain the effects of particle size, temperature, and stirring on a solid in solution?

_____ Can you explain the effect of temperature, stirring, and pressure on a gas in solution?

How soluble are two salts at varying temperatures?

Chapter 21

Perhaps the most familiar type of solution is a solid dissolved in water. When you add lemonade mix to water, you make lemonade, a water solution. No chemical change takes place when a solid is dissolved in a liquid. If the liquid evaporates, the original solid remains and it is chemically unchanged.

The maximum amount of a solute that can be dissolved in a solvent is called the solubility of the solute. Solubility of a solid is often expressed as the maximum number of grams of a substance that will dissolve in 100 g of solvent. The solubility of a substance is not the same under all conditions. For example, temperature changes can affect the solubility of a solid in water.

Strategy

You will determine the solubility of two salts.
You will determine the effect of temperature on the solubility of a salt.
You will interpret information from a solubility graph.

Materials

graduated cylinder (10-mL)   pot holder
beaker (250-mL)   4 small aluminum pie pans
hot plate   metric balance
thermometer   sodium chloride, NaCl(*cr*) (2 5-g samples per group)
4 test tubes   potassium bromide, KBr
test-tube rack   tap water
test-tube holder   distilled water

WARNING: *KBr is a body tissue irritant. Handle the thermometer carefully. Do not stir with the thermometer. If it breaks, do not touch anything. Inform your teacher immediately.*

Procedure

1. Safety goggles and a laboratory apron should be worn throughout this experiment. Pour tap water into the beaker until it is about one-third full. Heat the water on the hot plate until the temperature reaches 55°C–60°C. Use the thermometer to determine the temperature.

2. Label the four test tubes A, B, C, and D. Label the four aluminum pans A, B, C, and D. Find the mass of each pan and record it Table 1.

3. Get the four 5-g salt samples from your teacher. Add 5 g of NaCl to each of tubes A and B. Add 5 g of KBr to each of tubes C and D.

4. Using the graduated cylinder, add 5 mL of distilled water to each of tubes A through D.

Shake each tube to dissolve the salt, but be careful to avoid spilling the solution.

5. Carefully place tubes A and C in the water in the beaker and allow the contents to reach the temperature of the water bath, which will take about 5 min. Use the test-tube holder to remove the hot tubes to the test-tube rack. **WARNING:** *The tubes will be hot.*

6. Allow the four tubes to stand in the test-tube rack for a few minutes to allow any solid material to settle.

7. Using the test-tube holder, carefully pour the liquid from tube A into pan A. Do not transfer any of the solid. You will need to pour the liquid slowly. See Figure 1. Pour the liquids from the remaining tubes into the pans in the same way.

Laboratory Activity 2 (continued)

Figure 1

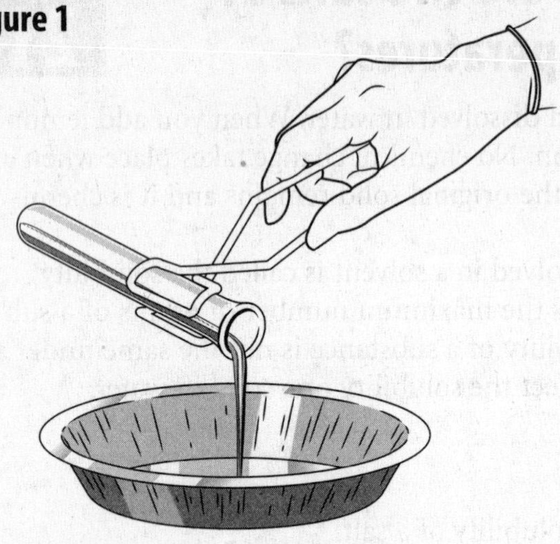

10. Determine the mass of the liquid evaporated from each pan by subtracting the mass of the pan and salt after evaporation from the mass of the pan, liquid, and salt. Record this information in the table.

11. Determine the mass of salt left in each pan after evaporation by subtracting the mass of the empty pan from the mass of the pan and salt. Record this information in the table.

12. Use the masses of the dissolved salts to determine the solubility per 100 g of water. Use a proportion in your calculations. Record the solubility in the table.

8. Determine the mass of each pan and its liquid. Record the masses in the table in the Data and Observations section.

9. Heat the pans on a hot plate using low heat. When all of the liquid evaporates, use a pot holder to remove the pans from the heat. **CAUTION:** *Do not touch the hot pans or the hot plate.* After the pans have cooled, find the mass of each and record this information in the table.

Data and Observations

Object being massed	Mass (g)			
	A	B	C	D
Empty pan				
Pan liquid, and salt				
Pan and salt				
Liquid evaporated				
Salt after evaporation				
Solubility				

Laboratory Activity 2 (continued)

Questions and Conclusions

1. What type of solid material settled to the bottom of each test tube?

_____

_____

2. Which salt had the greater solubility at 55°C–60°C?

_____

3. What would you expect to happen to the solubility of each salt if the temperature of the water were increased to 75°C?

_____

4. Look at the solubility graph in Figure 2. This graph shows how change in temperature affects the solubility of four common compounds.

Figure 2

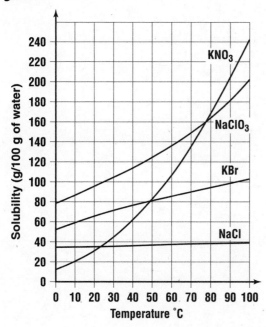

a. How does an increase in the temperature affect the solubility of NaCl?

_____

_____

b. How does an increase in temperature affect the solubility of KBr?

_____

_____

Laboratory Activity 2 (continued)

5. Refer to Figure 2. At what temperature does KNO₃ have the same solubility as KBr? What is the solubility at this temperature?

_____

_____

Strategy Check

_____ Can you determine the solubility of NaCl and KBr?

_____ Can you determine how temperature affected the solubility of NaCl and KBr?

_____ Can you interpret information from a solubility graph?

Acid Rain

Have you ever seen stained buildings, crumbling statues, or trees that have lost their leaves because of acid rain? Acid rain is a harmful form of pollution. Its effects are also easy to see. Acid rain is precipitation that contains high concentrations of acids. The precipitation may be in the form of rain, snow, sleet, or fog.

The major products formed from burning fossil fuels such as coal and gasoline are carbon dioxide and water. However, nitrogen dioxide and sulfur dioxide are also formed. These gases dissolve in precipitation to form acid rain.

When acid rain falls on a pond or lake, the acidity of the water increases. The rise in the acidity is usually harmful to organisms living in the water. If the acidity becomes too high, all living things in the water will die. The pond or lake is then considered to be "dead."

Strategy

You will generate a gas that represents acid rain.
You will observe the reaction of this gas with water.
You will demonstrate how the gas can spread from one location to another.

Equipment

96-well microplate
plastic microtip pipette
distilled water
paper towel
universal indicator solution

forceps
calcium carbonate, $CaCO_3(cr)$
scissors
soda straw
sealable, plastic sandwich bag

white paper
hydrochloric acid solution,
    $HCl(aq)$
watch or clock

WARNING: *The hydrochloric acid solution is corrosive. The universal indicator solution can cause stains. Avoid contacting these solutions with your skin or clothing. Wear an apron and goggles during this experiment.*

Procedure

1. Place the microplate on a flat surface.
2. Using the plastic microtip pipette, completely fill all the wells except A1, A12, D6, H1, and H12 with distilled water.
3. Use a paper towel to wipe away any water on the surface of the microplate.
4. Using the microtip pipette, add 1 drop of the indicator solution to each well containing water. Rinse the microtip pipette with distilled water.
5. Use the forceps to add a small lump of calcium carbonate to well D6.
6. Use the scissors to cut four 1-cm lengths of soda straw. Insert one length of soda straw in each of the wells A1, A12, H1, and H12 as shown in Figure 1. Cut a 0.5-cm length of soda straw and place it in well D6.
7. Carefully place the microplate into the plastic sandwich bag and seal the bag. Place the bag on the piece of white paper.

8. Using the scissors, punch a small hole in the plastic bag directly over well D6.
9. Fill the microtip pipette one-fourth full with the hydrochloric acid solution.
10. Slip the tip of the pipette through the hole above well D6. Direct the stem of the pipette into the soda straw in well D6.
11. Add 4 drops of hydrochloric acid to the well. Observe the surrounding wells.
12. After 30 seconds, note any color changes in the surrounding wells. Record a color change in the solution in a well by marking a positive sign (+) in the corresponding well of the microplate shown in Figure 2a in Data and Observations.
13. Repeat steps 11 and 12 two more times. Record your two sets of observations in Figure 2b and Figure 2c, respectively.

Laboratory Activity 2 (continued)

Figure 1

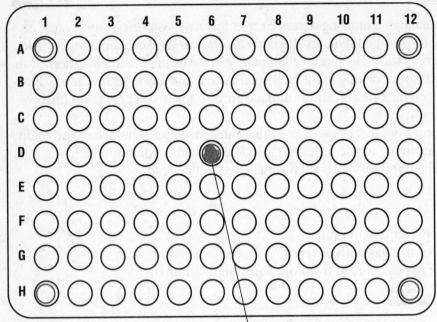

Calcium carbonate

Data and Observations

Figure 2a

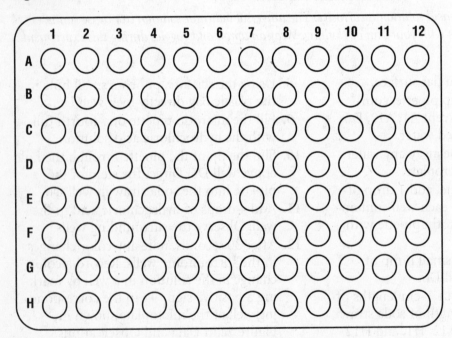

Laboratory Activity 2 (continued)

Figure 2b

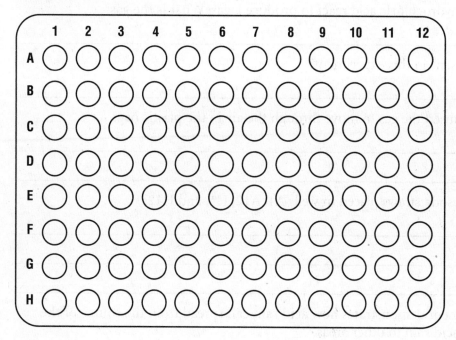

Figure 2c

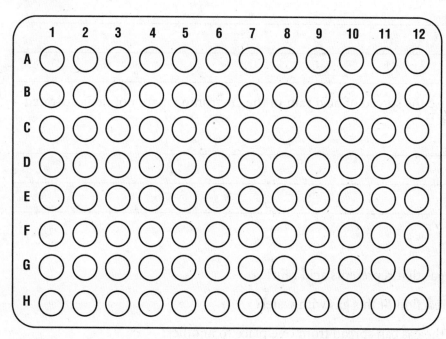

Laboratory Activity 2 (continued)

Questions and Conclusions

1. Calcium carbonate and hydrochloric acid react to produce a gas. What is the gas?

_____

2. What does this gas represent in this experiment?

_____

3. What physical process caused the gas to move through the air in the plastic bag?

_____

_____

4. Why were the lengths of soda straws placed in wells A1, A12, H1, and H12?

_____

_____

_____

5. Discuss how this experiment demonstrates how acid rain can spread from the source of the chemicals that produce acid rain to other areas.

_____

_____

_____

_____

6. What factors that may cause the spread of acid rain in the environment are not demonstrated in this model experiment?

_____

_____

_____

Strategy Check

_____ Can you generate a gas that represents acid rain?

_____ Can you detect the reaction of this gas with water?

_____ Can you show how the gas can spread from one place to another?

The Breakdown of Starch

LAB 1 Laboratory Activity

Chapter 23

Living things are made of carbon compounds called organic compounds. Many organic compounds are long molecules called polymers, which consist of small repeating units. Starch is a polymer of sugar units.

When you eat a piece of bread, your body breaks down the starch present in the bread. Substances in your saliva begin splitting the long starch polymers into shorter chains of sugar units. Digestion continues in your stomach and intestines until the shorter chains are broken down into individual sugar molecules. Finally, the sugar molecules combine with oxygen inside the cells to produce carbon dioxide and water and release energy. This energy allows you to run, stay warm, talk, think, and so on.

Strategy

You will use indicators to test unknown solutions for starch and sugar.
You will use a solution of saliva substitute to detect the breakdown of starch.

Materials

4 test tubes
solution X
solution Y
test-tube holder
starch indicator solution

sugar indicator solution
test-tube rack
250-mL beaker
thermometer
100-mL beaker

saliva substitute
4 rubber stoppers for test tubes
watch or clock

WARNING: *Starch indicator solution and sugar indicator solution are poisonous. Handle with care.*

Procedure

Part A—Starch and Sugar Indicators

1. Label the four test tubes A through D. Look at Table 1. Add 10 drops of unknown solution as indicated in Table 1 to the corresponding test tube. See Figure 1.

Figure 1

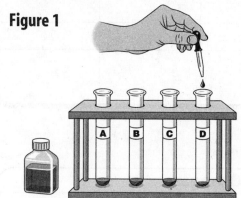

2. Use the test-tube holder to place tubes B and D in the boiling water bath provided by your teacher.
3. Add 1 drop of starch indicator solution to tubes A and C. If the indicator changes color, starch is present.

4. Record the color change in Table 1 and indicate which solution contains starch.
5. Add 1 drop of sugar indicator solution to tubes B and D and boil for 3 minutes. **WARNING:** *Tubes will be hot.* If the indicator changes color, sugar is present. Record the color changes in Table 1, and indicate which solution contains sugar. Using the test-tube holder, remove tubes B and D from the boiling water bath and place them in the rack to cool.
6. Rinse the test tubes with water.

Part B—Breakdown of Starch

1. Prepare a warm water bath in a 250-mL beaker. Make a mixture of warm and cool water to fill the beaker about half full. Use the thermometer to determine the temperature of the water. Add small amounts of warm or cool water using the 100-mL beaker until the temperature of the water bath reaches 35°C–40°C.

Laboratory Activity 1 (continued)

2. Use the test tube labeled A from Part A. Fill the test tube about 1/4 full with a solution of saliva substitute.

3. You tested solutions X and Y to determine which contained starch. Add 15 drops of the starch solution to the saliva substitute solution in the test tube. Stopper and shake the tube to mix the liquids. Note the time. Remove the stopper and place the tube in the warm water bath. See Figure 2.

4. After 5 minutes, pour two small and equal portions of the liquid in tube A into tubes B and C. Do not use all the liquid from tube A.

5. Using the procedure in Part A, not including step 1, place tube C in the boiling water bath. Test the liquid in tube B for the presence of starch with one drop of starch indicator solution. Test the liquid in Tube C with a drop of sugar indicator solution. Record your observations of color changes in Table 2. Continue to time the reaction in tube A.

Figure 2

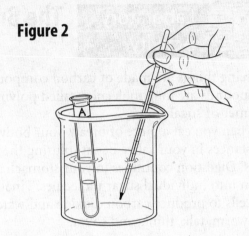

6. Leave tube A in the warm water bath. Add warm or cool water to the bath to adjust the bath temperature to 35°C–40°C.

7. Rinse tubes B and C with water.

8. After another 5 minutes, repeat steps 4–6. If time allows, repeat these steps a total of three times. Record your observations each time in Table 2.

Data and Observations

Part A—Starch and Sugar Indicators

Table 1

Tube	Unknown solution	Indicator solution	Color change	Starch (X)	Sugar (X)
A	X	starch			
B	X	sugar			
C	Y	starch			
D	Y	sugar			

Part B—Breakdown of Starch

Table 2

| Time (min) | Color Changes | |
	Starch indicator—tube B	Sugar indicator—tube C
5		
10		
15		

Laboratory Activity 1 (continued)

Questions and Conclusions

1. What happened to the starch solution when added to the saliva substitute? How do you know this?

   _____

   _____

   _____

   _____

2. Why is a water bath at a temperature between 35°C and 40°C used in this experiment?

   _____

   _____

   _____

3. If you chew a plain cracker without adding sugar, you will probably detect a sweet taste. Why does this happen?

   _____

   _____

   _____

   _____

4. Can the indicator solutions used in this experiment be used to determine how much sugar or starch is present in a sample? If so, explain how this could be done. If not, explain why this cannot be done.

   _____

   _____

   _____

Strategy Check

_____ Can you test unknown solutions for starch and sugar?

_____ Can you detect the breakdown of starch?

Testing for a Vitamin

Vitamin C is a complex organic compound found in many foods, such as fruits and vegetables. You can identify vitamin C by its chemical properties. It reacts with certain indicator solutions, causing the solutions to change color. The color change of the solution indicates that the vitamin C in the solution has reacted. You can determine the relative amounts of vitamin C in different foods by testing the food with the indicator solutions.

Strategy

You will observe the reactions of various concentrations of vitamin C with a color indicator.
You will compare the relative amounts of vitamin C in different types of orange juice and
orange drink.

Materials

96-well microplate

vitamin C indicator

sheet of white paper

freshly squeezed orange juice

plastic microtip pipette

bottled orange juice

vitamin C solution

orange drink

distilled water

WARNING: *The vitamin C indicator can cause stains. Avoid contacting it with your skin or clothing.*

Procedure

Part A—Testing the Vitamin C Solution

1. Place the 96-well microplate on a piece of white paper on a flat surface. Have the numbered columns of the microplate at the top and the lettered rows at the left.

2. Using the microtip pipette , add 10 drops of the vitamin C solution to well A1. Rinse the pipette with distilled water.

3. Use the pipette to add 5 drops of distilled water to each of the wells A2–A6.

4. Remove most of the solution from well A1 using the pipette. Add 5 drops of this solution to well A2. Return the solution remaining in the pipette to well A1. Rinse the pipette with distilled water.

5. Use the pipette to mix and then remove most of the contents of well A2. Add 5 drops of this solution to well A3. Return the solution remaining in the pipette to well A2. Rinse the pipette with distilled water.

6. Repeat step 5 for wells A4–A6 as shown in Figure 1.

Figure 1

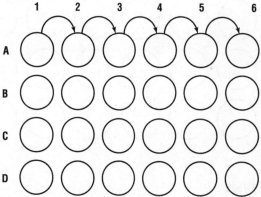

7. Using the pipette, add 3 drops of vitamin C indicator solution to each of the wells A1–A6. Stir the contents of each well with a clean toothpick.

8. Observe the color of each well. If you see a color change, mark a positive sign (+) in the corresponding well of the microplate shown in Figure 2 in the Data and Observations section. Record no change in color as a zero (0).

9. Rinse the pipette with distilled water.

Laboratory Activity 2 (continued)

Part B—Testing Orange Juices for Vitamin C

1. Add 10 drops of freshly squeezed orange juice to well B1 using the pipette. Rinse the pipette with distilled water.
2. Add 5 drops of distilled water to each of the wells B2–B6.
3. Mix and then remove most of the contents of well B2 using the pipette. Add 5 drops of this solution to well B3. Return the solution remaining in the pipette to well B2. Rinse the pipette with distilled water.
4. Repeat step 3 for wells B3–B6.

5. Add 3 drops of the vitamin C indicator solution to each of the wells B1–B6. Stir the contents of each well with a clean toothpick.
6. Observe the color of each well. Record your observations of wells B1–B6 by marking a + or 0 in the corresponding wells shown in Figure 2.
7. Rinse the pipette with distilled water.
8. Repeat steps 1–7 for the bottled orange juice and the orange drink. Place the orange juice in well C1 and the orange drink in well D1.

Data and Observations

Figure 2

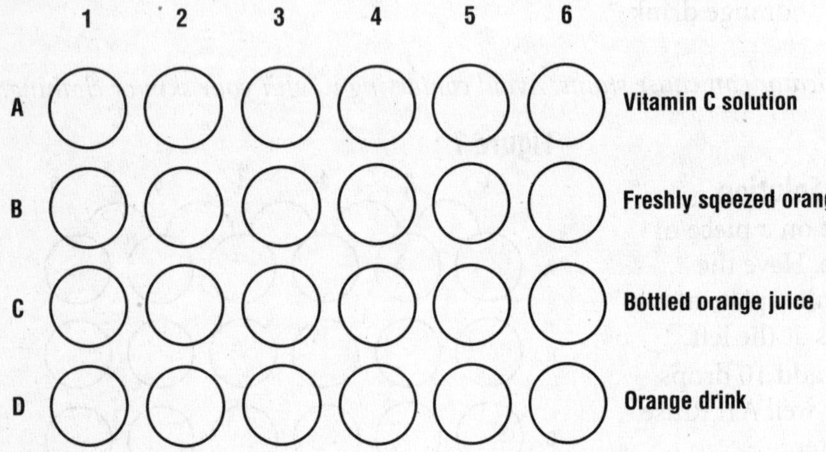

Row A: Vitamin C solution
Row B: Freshly sqeezed orange juice
Row C: Bottled orange juice
Row D: Orange drink

Columns: 1 2 3 4 5 6

Analysis

A color change indicates that vitamin C is present in the solution.

Questions and Conclusions

1. What happens to the concentration of the solutions in the wells as one moves across a row?

2. Which solution had the greatest concentration of vitamin C? How do you know?

Laboratory Activity 2 (continued)

3. Which of the three food products tested had the greatest concentration of vitamin C?

_____

_____

4. A sample of freshly squeezed orange juice contains 5.0 mg of vitamin C in 15 mL of the juice. How much juice must you drink to meet a daily recommended requirement of 32 mg of vitamin C?

_____

5. Vitamin C reacts with oxygen in the air. How does the length of time orange juice is exposed to air affect the amount of vitamin C in it? Design an experiment to answer this question.

_____

_____

_____

Strategy Check

_____ Can you determine the concentration of vitamin C in a substance?

_____ Can you determine the amounts of vitamin C in different types of orange juice and orange drink?

Soldering with an Alloy

Chapter 24

Solder (SOD er) is a very useful household alloy. It is used to patch or join metals in items ranging from jewelry and crafts to metal pipes to electrical circuits and broken wires. It is inexpensive and easy to find in hardware stores. Its relatively low melting point and quick solidifying make it easy to use. A soldering tool provides the heat to melt the solder. This tool is usually a special rod of metal that is attached to an electrical wire. The tip of the soldering tool becomes heated according to the proper temperature for the type of solder that is used. Hard solders contain combinations of silver, copper, and zinc. Soft solders are the most common and contain between 30 to 70 percent tin with other compounds, usually lead. In this lab you will examine a tin solder and make an electrical connection with it.

Strategy

You will observe and describe the properties of tin solder.
You will use solder to connect copper wires.

Materials

copper wire cut into two 15-cm lengths
tin solder wire (about 15 cm long)
soldering tool

ceramic tile (large one for placing soldering tool
   and wires on)
aluminum foil

Procedure

1. Place students into groups of three or four. Examine the properties of the solder wire. Look for ductility, malleability, color, and luster. Refer to the information given in the Data and Observations section. Choose a number to represent the property and record your observations in the table provided.

2. Repeat the same procedure for the copper wire.

3. Take a tile and completely cover it in aluminum foil. Ceramic tile is used as a non-conducting type of material. Completely cover the tile with aluminum to prevent any soldering from eventually sticking to it.

4. Plug in your soldering tool and set it to heat up to tin soldering. If you have the type of soldering tool with no temperature settings, just let it heat up for about five minutes. **WARNING:** *Do not touch the metal end of the tool.* The colored plastic end is made so that it will conduct heat and is safe to hold.

5. Place your soldering tool on the covered tile and wait for it to heat. **WARNING:** *Point the hot tip away from any papers or flammable materials. It will burn a hole through clothing so remove all jackets and backpacks from the lab table.*

6. Collect two pieces of copper wire and place them end to end on your tile. Have one member of the group hold the wires in place. Have another hold the length of solder at the exact place where the two copper wires meet. The point is to make the solder a liquid metal for just a second so that it will bond the two pieces of copper wire together.

7. A third student should just barely touch the tip of the tool to the place where the solder meets the copper wire. It may take some practice getting the solder to melt and stick to the copper wire. Professional electricians practice for years before they make small and efficient solder connections.

8. When you have successfully soldered the two pieces of copper wire together, show the finished product to your teacher. You now have a wire that will conduct electricity and is twice as long as when you started. You have effectively used an alloy for repair purposes.

Laboratory Activity 1 (continued)

Data and Observations

Malleability:
1 – 2 = not malleable
3 – 4 = slightly malleable
5 – 6 = malleable
7 – 8 = very malleable
9 – 10 = highly malleable (unable to keep a shape for very long)

Ductility:
1 – 2 = not ductile
3 – 4 = slightly ductile
5 – 6 = ductile
7 – 8 = very ductile
9 – 10 = highly ductile

Luster:
1 – 2 = not shiny
3 – 4 = slightly shiny
5 – 6 = shiny
7 – 8 = very shiny
9 – 10 = highly shiny

Metal or alloy	Color	Malleability	Ductility	Luster
1. Tin solder				
2. Copper wire				

Questions and Conclusions

1. Why is the solder called an alloy and the copper wire is not?

_____

_____

2. What was the source of the energy for changing the state of the solder?

_____

_____

Laboratory Activity 1 (continued)

3. Why was solder chosen to join the two copper wires together and not something like clay or glue?

_____

_____

_____

4. What happened when you tried to make your first soldering joint?

_____

_____

_____

5. How was the solder different from the copper wire when the finished product was made?

_____

_____

_____

_____

6. Why don't electricians just use solder wire instead of copper wire to make electrical connections?

_____

_____

_____

_____

Strategy Check

_____ Can you observe and describe the properties of tin solder?

_____ Can you use solder to connect copper wires?

Investigating Polymers

Polymers are very large molecules made of many identical units linked together in a chain. The individual units, called monomers, can be very simple or rather complex. A single molecule of a polymer can contain tens of thousands of monomers. The identity of the monomers, the way they are linked, and the length of the chains are some of the characteristics that determine what the polymer looks like and how it can be used. In some polymers, the monomers link together end to end, resembling a string of identical beads. Some monomers contain side chains, or branches, and resemble a very long string of large paper clips attached in a chain with chains of small paper clips hooked on in just the same way at just the same interval. In some polymers, the way the monomers link together causes the polymer molecule to bend and twist and sometimes form crosslinks in the long chains. Some polymers, such as polyethylene, contain just carbon and hydrogen with carbon forming the backbone of the molecules. Other polymers contain elements other than carbon and hydrogen. Nylon contains nitrogen, and polyvinyl chloride (PVC) contains chlorine. Some polymers, such as the polymer that is sold as toy putty, have a silicone backbone rather than a carbon backbone.

Every difference in molecular structure holds the possibility that the polymer can be used in a different way. As chain length increases, physical properties of the polymer change. Polymers made from longer chains are more ductile and are harder than polymers with shorter chains. Longer chain molecules are more viscous than shorter chain molecules. More crosslinking within a polymer molecule usually results in a more rigid substance.

Strategy

You will observe properties of different polymers.
You will describe some properties of various polymers.

Materials

variety of samples of polymers
newspaper
waxed paper
small paper cups

epoxy cement
large toothpicks
*wooden craft sticks
permanent adhesive

rubber cement
bathroom caulking
toy putty
*Alternate materials

Procedure

Part A—Observing Physical Properties

1. Spread at least two or three sheets of newspaper over the surface of your lab table to protect the tabletop. Place a small sheet of waxed paper on top of the newspaper.

2. Be sure to wear gloves when handling any of the materials in this experiment. Examine the properties of at least six polymers. List the names of the materials selected under *Materials* in Table 1 in the Data and Observations section. For polymers like caulking that are in containers, use a toothpick to put a small sample on the waxed paper.

Observe the color and odor of each. Record your observations in Table 1 in the Data and Observations section. Note how thick the substances are (the thicker the substance, the higher the viscosity) and record your observation under *Viscosity* in Table 1. Observe how hard or brittle the solids are. Record your observations in Table 1.

3. Put the waxed paper with the samples aside.

Laboratory Activity 2 (continued)

Part B—Comparing Physical Properties

1. Cover the newspaper with another sheet of waxed paper.
2. Be sure to wear gloves when handling any of the materials in this experiment. Following the instructions your teacher gives, mix a small amount of epoxy glue in a small paper cup. Use a toothpick to stir the epoxy. Avoid getting any epoxy glue on your hands. Pour some of the mixed epoxy on the waxed paper. Use a toothpick to spread the epoxy into a thin sheet.
3. Using separate toothpicks, spread small amounts of the permanent adhesive, the toy putty, the rubber cement, and the bathroom caulking into thin sheets at various places on the wax paper. You may have to apply a second, and perhaps a third, coat of rubber cement. Observe these samples for color and odor. Using a toothpick, observe the viscosity, and brittleness or hardness of the samples. Do not touch the samples with your fingers. Record your observations in Table 2.

4. Allow about 10 minutes for the polymers to dry. Test the slower drying polymers for dryness by poking them with a toothpick to see if any tackiness, or stickiness, remains. Wait to make your observations until the samples are dry.
5. When your samples are dry, once again observe them for color, odor, viscosity, brittleness, and hardness. Record your observations in Table 3 in the Data and Observations section.
6. Remove the thin sheets of dried polymers from the waxed paper. Under *Adhesion* in Table 3, note whether the polymer came off the waxed paper easily.

Data and Observations

Table 1

Material	Color	Odor	Viscosity/Brittleness/Hardness

Laboratory Activity 2 (continued)

Table 2

Initial Observations			
Material	**Color**	**Odor**	**Viscosity/Brittleness/Hardness**

Table 3

Final Observations				
Material	**Color**	**Odor**	**Viscosity/ Brittleness/ Hardness**	**Adhesion**

Questions and Conclusions

1. Describe the range of physical properties, including color, odor, viscosity, brittleness, and hardness in the six samples you selected in Part A.

_____

_____

_____

_____

_____

Laboratory Activity 2 (continued)

2. Describe some of the uses that the polymer samples have.

_____

_____

_____

3. Was the rubber cement or the epoxy glue harder when it dried? Which would you expect or to have more crosslinks in its molecules? Explain your answer.

_____

_____

_____

4. Would you expect the molecules that form a plastic bag to have longer or shorter chains than the molecules that form a milk carton? Explain your answer.

_____

_____

Strategy Check

_____ Can you observe the properties of different polymers?

_____ Can you describe some properties of various polymers?

Tasty Quake

To understand Earth processes that can't be seen, it helps to have a simulation, or model, that can be observed. Scientists create simulations using a variety of techniques, from the most simple to complex computer programs.

Scientists know that the S-waves, or surface waves, generated by earthquakes cause the most damage to artificial structures. This is because S-waves produce lateral motion. Most artificial structures are not designed to withstand such sideways motion, although in earthquake-prone areas, architects are making efforts to design buildings to make them less likely to collapse during an earthquake.

Strategy

You will demonstrate the effect of earthquake waves on Earth's surface and man-made structures.

Materials

pan-prepared gelatin dessert
clear plastic wrap
sugar cubes, dominoes, or dice

Procedure

1. Prepare your gelatin: empty two 6-oz boxes of gelatin dessert and two 1-serving envelopes of unflavored gelatin into a 9 × 12 metal baking pan. Add four cups of boiling water. Stir to mix, then chill in refrigerator until set.
2. When the gelatin has set, hold the pan with one hand and gently tap the side.
3. Observe what happens to the gelatin.
4. Repeat step 2 several times, tapping harder each time.
5. Observe what happens after each tap. Record your observations in the Data and Observations section.
6. Cover the gelatin with plastic wrap, making sure the wrap rests on the gelatin.
7. Use sugar cubes, dominoes or dice to create a "building" on the plastic wrap.
8. Tap the pan again.
9. Observe what happens to your miniature building. Record your observations and estimate the value of the simulated earthquake on the Modified Mercalli Intensity Scale found in your text based on the state of your miniature building and the movement visible during the simulation.
10. Reassemble your structure and repeat step 9 several times, tapping harder each time.

Laboratory Activity 1 (continued)

Data and Observations

Trial	Observations	Mercalli Index
1		
2		
3		
4		
5		

Questions and Conclusions

1. **Evaluate** how the gelatin reacted to taps of various strengths.

_____

_____

2. **Infer** how the reaction of the gelatin to the taps is similar to the way Earth's surface reacts to earthquake waves.

_____

_____

3. **Describe** the effect the waves moving through the gelatin had on your miniature structure.

_____

_____

4. **Compare** your observations and Mercalli Index values to those of your lab partners. Infer why we use the Richter Scale today instead of the Mercalli Intensity Scale.

_____

_____

5. **Explain** what could be changed with your structures to reduce the damage due to an earthquake.

_____

_____

Strategy Check

_____ Can you effectively demonstrate the effect of earthquake waves on Earth's surface and man-made structures?

Tsunami in a Box

Much of the damage caused by earthquakes actually comes from an earthquake's secondary effects, such as fires, floods, or tsunamis. Tsunamis, also known as tidal waves, are generated by a sudden displacement of water. It usually takes an earthquake with a magnitude greater than seven on the Richter scale to generate a significant tsunami. Tsunamis travel from their point of origin at a speed of 300–350 miles per hour. Earthquake waves travel 50 times faster than tsunamis, so seismographs can give people time to prepare for the arrival of a tsunami after an earthquake has occurred.

Strategy

You will create a model of the effects of a tsunami.

Materials

glass or metal baking pan or plastic shoe box
1 L of water
plastic lid from coffee or margarine container
punching tool or drawing compass
scissors
string
sand
various small objects to represent shoreline features
book or block of wood to serve as wedge
metric ruler

Procedure

1. Use the wedge to tilt the box or pan to an angle of about 20°.
2. Pour water into the box or pan to cover the lower end, leaving about a third of the box or pan dry at the upper end.
3. Pack a layer of sand 2-3 cm thick on the dry end of the container to simulate a beach. Use your hands to mold dunes or drifts. Draw roads parallel to the shore. Build docks and other small, lightweight structures to complete the shore environment.
4. Sketch an overhead view of the structures in the sand, including the edge of the beach. Use the ruler to make this layout as accurate as possible.
5. Punch a hole in the plastic lid near the rim and thread it with a string 20 cm long. Tie knots to hold the string in place.
6. Gently and without making waves place the plastic into the bottom of the water at the deep end of the box or pan. Trim to fit if necessary. The string should hang over the low side of the box or pan.
7. Firmly hold the lid down where it faces the shallow water and, using your other hand, pull the string at the deep end up with a rapid movement.
8. Observe and record the results in the Data and Observations section.
9. Sketch an overhead view of the structures left after the wave has receded. Use the ruler to measure the new edge of the beach.

Laboratory Activity 2 (continued)

Data and Observations

_____

_____

_____

_____

Questions and Conclusions

1. **Evaluate** how the sudden rush of water affected your miniature shoreline.

_____

_____

2. **Infer** what the sudden motion of the plastic lid represented.

_____

_____

3. **Infer** what different effects various strengths and motions of the plastic lid have on the shoreline.

_____

_____

4. **Describe** what could be done in the sand to protect structures from the incoming water.

_____

_____

5. **Infer** what effect increasing the distance from the plastic lid to the shoreline has on the resulting tsunami.

_____

_____

Strategy Check

_____ Can you create a realistic model of a tsunami?

The Colored Ribbons of Gneiss

Gneiss is a high-grade metamorphic rock that originates from either sedimentary rock (usually shale) or from the igneous rock granite. It is often used as building or paving stone. Gneiss is labeled high-grade because it was subjected to more heat and compression than schist, the first stage of the metamorphism of shale. As a result, it is coarser and has distinct color layers, or banding.

You may remember that color variation within sedimentary rock is due to the difference in particle size of the various sediments. However, particle size is no longer a factor in metamorphic rock because intense heat and pressure have made particle size more uniform. The color banding in the metamorphic rock gneiss, therefore, comes solely from the difference in mineral colors. Gneiss is usually composed of mica, feldspar, and quartz, although it can also contain hornblende, kyanite, garnet, tourmaline, magnetite, and other minerals.

Glassy in appearance, quartz can appear white, gray, yellow, or even red in sedimentary rock, but it is gray in igneous rock. Glassy-faced potassic feldspars—microline and orthoclase—are pink or tan in appearance, although occasionally they look white. Another type of feldspar, plagioclase feldspar, ranges from white to dark grey (sometimes black) and may show small flashes of green or blue. The plagioclase feldspars are albite and labradorite.

The micas (muscovite and biotite) are thin flakes or flake layers, that peel easily off a rock sample. Muscovite ranges from silver to brown, while biotite is black. Other "flaky" minerals are the chlorite group, although their flakes aren't as thin as the mica group's. Chlorite always has a green cast and varies from medium to dark green, although it can also be black with a green tinge.

Although gneiss is usually light-colored, it can also appear dark. A rock hunter who observes fairly broad color stripes in a coarse-textured, hard rock can be fairly certain that the rock is gneiss. However, if this gneiss sample is compared with another, the difference in coloration can be dramatic.

In this experiment, you will create "gneiss," observe banding, identify the "minerals" that create the banded colors, and observe the different combinations that can make gneiss rocks differ greatly from one another. Predict how your model will compare with those of your classmates. Record your prediction and observations in the Data and Observations below.

Strategy
You will create a model of the metamorphic rock gneiss.
You will observe mineral banding in the model.
You will hypothesize about specific minerals that contribute to banding.
You will compare and contrast your model with those of other students.

Materials
sheet of newspaper
wide-mouthed canning jar, 1-quart or 1-L (must be clear and colorless)
granulated sugar (not extra-fine)
colored sugar (yellow, red, green, blue, pink, black)
clean, dry beach sand
small scoop
*colored sand (such as that used in candle-making)
*small serving spoon
*Alternate materials

Laboratory Activity 1 (continued)

Safety Precautions

WARNING: *Students should not eat the sugars in this activity.*

Procedure

1. Place a sheet of newspaper on your work surface.
2. Obtain "mineral" samples from your teacher.
3. Place enough sand in the bottom of the container to form a half-inch (about 1.3 cm) layer.
   Do **not** shake the container to level the sand.
4. Add one color of sugar to the container in an amount that will form an observable layer.
   Do **not** shake the container to level the sand.
5. Continue to create observable bands by adding other sugar colors, periodically alternating with
   an observable sand layer. (This sand layer does not have to be as deep as the initial layer.)
   Do **not** shake the container between additions.

Data and Observations

1. **Predict** how the appearance of your gneiss model will compare to that of your classmates'
   models. Explain your prediction, based on the variables within the lab activity.

   _____

   _____

   _____

   _____

Mineral Color	Possible Mineral Identification

Laboratory Activity 1 (continued)

Questions and Conclusions

1. Why was it important to allow the bands to form without leveling each layer?

_____

_____

_____

2. After observing the colored bands in the "gneiss" model, **infer** one reason why gneiss is a popular construction and paving material, aside from its hardness property.

_____

_____

_____

3. Explain why the comparison of your model to those of your classmates might help you to identify gneiss in the future.

_____

_____

_____

Strategy Check

_____ Can you identify gneiss based on its physical appearance?

_____ Can you predict why one gneiss sample might differ from another?

LAB
2 Laboratory Activity

Making Sedimentary Rocks

Chapter 26

As you've learned, there are three types of sedimentary rock: clastic (such as sandstone), chemical (such as limestone), and biochemical, or organic (such as soft coal). Although each of these rock types is formed from the deposition of sediment, there are differences in the formation process that affect the physical properties of each type.

Most sedimentary rocks are clastic. Other names for clastic are detrital or terrigenous. Clastic rock originates from the weathering of existing rock. Rock bits and pieces separate and are transported to an area called the depositional basin, where they accumulate and are compacted into rock. Their clastic structure is broken, or fragmented, and consists of clasts (large pieces of gravel and sand), matrix (fine sand or mud that surrounds the clasts), and mineral cements (such as silica, iron oxide, and calcite). Clastic rocks are classified according to their clast size:

- Grain size greater than 2 mm, rounded— conglomerate (rough)
- Greater than 2 mm, angular— breccia (rough)
- 0.06 mm to 2 mm — sand (coarse)
- 0.002 mm to 0.06 mm — silt (gritty)
- Less than 0.002 mm — clay (smooth)

Chemical sedimentary rocks need water to supply dissolved minerals that are either evaporites (example, rock salt), carbonates (example, limestone), or silicates (example, chert).

Biochemical sedimentary rocks lack minerals but are mainly composed of decayed plant matter. An example of a biochemical rock is the soft coal lignite.

How does texture identify a sedimentary rock? Record your hypothesis in the Data and Observations section of this lab activity. After you have created three rock models, record your observations and identify each model as either *soft coal, limestone,* or *conglomerate* in the table below.

Strategy

You will model three types of sedimentary rock.
You will observe and describe their composition and texture.
You will identify the type of rock based on its composition and texture.

Materials

milk cartons, pint (3)
gravel, rounded
decomposing leaves
sand
water
soil
cement
plaster
water
mixing spoon

Safety Precautions

WARNING: *Avoid skin or clothing contact with moistened cement or plaster. Wear protective apron. Wash skin immediately if it comes in contact with mixtures containing plaster or cement.*

Laboratory Activity 2 (continued)

Procedure

1. Cut the tops off each of the milk cartons.

2. Label the cartons *A*, *B*, and *C*.

3. Add sand to carton A until it is half-full.

4. Stir enough water into the sand in carton A until the sand is wet.

5. Add 3 teaspoons of cement and several pieces of gravel to carton A.

6. Stir the mixture until the cement and gravel are well-distributed. Set carton A aside to dry.

7. Repeat steps 3 and 4 with carton B, substituting soil for sand.

8. Add three or four decomposing leaves as well as 3 teaspoons of cement to the soil in carton B.

9. Stir the mixture until the cement and leaves are well-distributed. Set carbon B aside to dry.

10. Add plaster to carton C until it is half-full.

11. Stir in enough water so that the plaster is moistened. (Mixture should not be runny.) Set carton C aside to dry.

12. Observe each carton after the mixtures have dried. Note observations in the Data and Observations table.

Data and Observations

Hypothesis:

_____

_____

_____

_____

Laboratory Activity 2 (continued)

Table 1

Carton	Composition & Texture	Rock Identification: Soft Coal, Limestone, or Conglomerate
A		
B		
C		

Questions and Conclusions

1. Which ingredient(s) in each rock made identification easy? **Explain**.

_____

_____

_____

2. **Explain** why sand was not added to carton C.

_____

_____

_____

3. Which rock is an example of a clastic sedimentary rock? Defend your answer.

_____

_____

_____

4. Which rock is an example of a biochemical rock? Defend your answer.

_____

_____

_____

Strategy Check

_____ Can you identify different types of sedimentary rock?

_____ Does texture help you to identify sedimentary rock?

Whatever the Weathering

In nature, matter is recycled. Chemical, physical, and biological processes form interconnected cycles that operate to change materials. Because we tend to study each process one at a time, you might get the impression that they stand alone. This is not true. For example, this activity involves two experiments.

The steel wool experiment might be performed by a chemist, since it involves chemical change. The soft and hard experiment might be performed by a geologist, since it involves physical change. If the two scientists do not share the results of their work, they will miss an opportunity to make important connections.

A real understanding of nature comes from learning how cycles interconnect. Weathering is an example of physical and chemical processes that involve the interaction of air, water, and rock over time.

Strategy
To model examples of both chemical and mechanical weathering

Materials
sugar cubes (10)
jar with a lid
gravel (10 pieces)
steel wool (2 small pieces)
small plastic bags (2)

Procedure

Part One: Steel Wool Experiment

1. Label one bag *dry* and put one small piece of steel wool in that bag.
2. Label the other bag *wet*. Moisten the other piece of steel wool and put it in that bag.
3. Both bags should be sealed and stored for three to four days.

Analysis

1. Examine the wool from the *dry* bag. Describe its appearance.

   _____

   _____

2. Work over a piece of white paper. Carefully roll the piece of dry steel wool between your fingers. Describe what has fallen on the white paper.

   _____

   _____

3. Examine the steel wool from the wet bag and repeat the steps above.

   _____

   _____

Laboratory Activity 1 (continued)

Questions and Conclusions

1. How were the wet and dry steel wool different?

_____

_____

2. What caused the changes in the wet steel wool?

_____

_____

3. What kind of weathering is this?

_____

_____

Procedure

Part Two: Soft Versus Hard

1. Place the sugar cubes in the jar and shake 20 times.
2. Pour the contents of the jar onto a piece of paper, separating the sugar cubes and the crumbs.
3. Return the sugar cubes to the jar and shake another 20 times.
4. Again pour the contents onto a piece of paper, separating the crumbs and the cubes.
5. Repeat the experiment using pieces of gravel instead of sugar cubes.

Data and Observations

Sugar cubes:

_____

_____

_____

_____

_____

Gravel:

_____

_____

_____

_____

_____

Laboratory Activity 1 (continued)

Analysis

1. Did the second shaking cause the cubes to look more worn?

_____

2. Is the amount of crumbs greater than, the same as, or less than the first shaking?

_____

3. How did the gravel react to the shaking compared to the sugar cubes?

_____

_____

Questions and Conclusions

1. Describe any changes you notice in the sugar cubes and the gravel.

_____

_____

2. Were these changes due to chemical or mechanical weathering? Why?

_____

_____

Strategy Check

_____ Did you model chemical and mechanical weathering?

Squeezing Water from a Well

Chapter 27

Artesian wells are formed when an aquifer is trapped between two aquitards. The aquitards prevent the water in the aquifer from flowing in any direction but up. The pressure applied by the aquitards and the constant addition of water from the saturation zone forces the water in the aquifer to the surface.

Strategy

To create a model of an artesian well

Materials

deep baking pan
medium–sized sponge
plywood about the same size as the sponge (2 pieces)

Procedure

1. Pour about two inches of water into the pan.
2. Wet the sponge and squeeze the water out until it is just moist.
3. Position the sponge between the two pieces of plywood, like a sandwich.
4. Position the sponge and wood so that one of its edges is immersed in the water in the pan. Make sure that the sponge and both pieces of wood are touching the bottom of the pan.
5. Observe what, if anything, happens for two minutes.
6. Squeeze the plywood pieces together to apply pressure to the sponge.
7. Observe what happens.
8. Leave the sponge and wood in place for another two minutes, then repeat steps 6 and 7.
9. Observe the results.

Data and Observations

Step 5:

_____

_____

Step 7:

_____

_____

Step 9:

_____

_____

Laboratory Activity 2 (continued)

Analysis

1. What does the water in the pan represent?

_____

2. What does the sponge represent?

_____

3. What do the pieces of plywood represent?

_____

4. What happened when you applied pressure to the sponge?

_____

Questions and Conclusions

1. **Describe** how the water entered the sponge.

_____

_____

2. **Infer** what would have happened if the top was the only exposed side of the sponge.

_____

_____

3. **Describe** what happened to the water that escaped from the top of the sponge.

_____

_____

Strategy Check

_____ Did you make a model of an artesian well?

Why Maritime Climates Are Mild

The seasonal temperature fluctuations in locations surrounded by land are much greater than the temperature fluctuations in locations surrounded by an ocean or near an ocean. The reason for this is that the specific heat of water is much greater than the specific heat of soil and rocks. This means it takes more heat energy to raise the temperature of one gram of water 1°C than one gram of soil or rock 1°C. This characteristic of a substance is called the specific heat of the substance.

Strategy

You will compare the specific heat of a substance with the specific heat of water.

Materials

balance scale with two trays
plastic foam cups (4)
thermometers with stands (2)
Bunsen burner with a tripod
glass beaker with tongs that can hold it
piece of metal
lids (2)

Procedure

1. Use the plastic foam cups to make two calorimeters by putting one cup inside the other. Use the thermometer stands to lower the thermometers into the cups. Place the two calorimeters on the balance scale, each about a third full with cool water. Adjust the amount of water, if need be, so that the calorimeters are balanced.

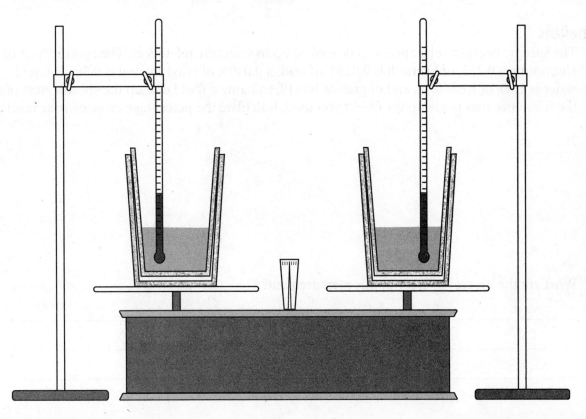

Laboratory Activity 1 (continued)

2. Record the two temperatures of the water in the calorimeter.
3. Place the piece of metal inside the beaker half-full with water. Use the Bunsen burner and tripod to bring the water and metal to near boiling point.
4. Using tongs, remove the object from the hot water and quickly place it in one of the calorimeters. Stir the water and measure the rise in temperature.
5. Using tongs, take the beaker of hot water and pour the hot water into the calorimeter without the object until the calorimeters balance again.

Data and Observations

	Calorimeter with Object	Calorimeter with Hot Water
Temperature Before		
Temperature After		

Calculations

The specific heat of the metal object is the change in the temperature (T) of the object divided by the change in the temperature of the water:

$$\text{specific heat of metal object} = \frac{T_{object}^{after} - T_{object}^{before}}{T_{water}^{after} - T_{water}^{before}}$$

Analysis

1. The specific heats of substances can be looked up in scientific references. The specific heat of aluminum is 0.215, of bismuth is 0.0294, of lead is 0.0305, of ethyl alcohol is 0.58, of liquid water is 1.00, of ice is 0.49, and of granite is 0.19, to name a few. Look up the specific heat of the substance that made up the object you used. Calculate the percentage error of your result.

2. What are the sources of error in this measurement?

_____

_____

_____

Laboratory Activity 1 (continued)

Questions and Conclusions

1. Is the mass of the added hot water equal to the mass of the metal object? How do you know?

   _____

   _____

2. Was the temperature of the added hot water the same as the temperature of the metal object before immersion? How do you know?

   _____

   _____

3. Did the hot water and the object give up the same amount of heat to the calorimeter? Give your reasons.

   _____

   _____

4. How does this experiment explain why climates around oceans have smaller changes of temperatures?

   _____

   _____

Strategy Check

_____ Do you know how to measure the specific heat of a substance?

_____ Do you understand why climates inland are not as temperate as climates near large bodies of water?

Measuring the Latent Heat of Vaporization

Radiant energy from the Sun melts ice and evaporates water. When water vapor condenses to form the water droplets that make up clouds, the heat required to vaporize the water is released into the atmosphere. In the same way, when liquid water solidifies, the latent heat of fusion (solidification) leaves the water and enters the surroundings.

Heat is measured in units of heat called calories. A calorie is defined as the amount of heat needed to raise the temperature of one gram of liquid water one degree (1 calorie = 1 g × 1°C). When water is heated up the temperature rises until it reaches 100°C. At this temperature, the heat changes the water from the liquid phase to the vapor phase. The amount of heat needed to transform one gram of liquid water at 100°C to one gram of gaseous water is the latent heat of vaporization. The accepted value for the latent heat of vaporization is 546 calories per gram.

Strategy

In this lab we will measure the heat output of a Bunsen burner and tripod setup per unit of time. Based on the amount of time it takes to vaporize a measured amount of water, we can calculate the latent heat of vaporization.

Materials

Bunsen burner
tripod
balance scale
thermometer
beaker

Procedure

1. Using a balance scale or a graduated cylinder, measure an amount of water and place it in a beaker. Measure the temperature of the water.
2. Adjust the flame of a Bunsen burner so that it delivers a steady amount of heat to the beaker of water. Heat the beaker to its boiling point. Record the time heating began and the time boiling began.
3. Boil the water for an additional length of time. The time should be long enough so that a measurable amount of water evaporates. Remove the heat and record the time.
4. Measure the mass of the remaining water in the beaker.

Data and Observations

Mass of initial amount of water (m_i)	
Temperature of the initial amount of water (T_i)	
Time heating began (t_i)	
Time boiling began (t_{boil})	
Time heating stopped (t_f)	
Mass of final amount of water (m_f)	

Laboratory Activity 2 (continued)

Calculations

The number of calories delivered to the water to make it boil is

$$Q = m_i \times (100°C - T_i)$$

The number of calories delivered per second is

$$R = \frac{Q}{t_{boil} - t_i}$$

The latent heat of vaporization is

$$L = \frac{R \times (t_f - t_{boil})}{m_i - m_f}$$

The percentage error is

$$error = \frac{546 \text{ cal/g} - L}{546 \text{ cal/g}}$$

Analysis

1. What are the possible sources of the error in your calculation?

_____

_____

_____

2. How could the experimental design be changed to improve the accuracy of the results?

_____

_____

_____

3. Before the water starts boiling, water is lost through evaporation. How would you determine if this loss of water affected the final result?

_____

_____

_____

Laboratory Activity 2 (continued)

Questions and Conclusions

1. Why does the temperature of boiling water being heated remain constant at 100°C?

_____

_____

2. When water vapor condenses to form clouds, does the temperature of the surrounding air increase or decrease? Explain your answer.

_____

_____

3. When solid glass is heated it gets softer until it becomes a liquid. Does ice get continually softer when it is heated?

_____

_____

_____

Strategy Check

_____ Do you know how to measure the latent heat of a substance?

_____ Do you understand why it requires energy to melt ice or evaporate water?

How to Measure the Circumference of Earth

LAB 1 Laboratory Activity

Eratosthenes, who lived in Alexandria around 250 B.C., measured the circumference of Earth by measuring the angle with which the Sun's rays strike the surface of Earth at different points. Science students around the world replicate this method by measuring the acute angle made by the intersection of the rays of the Sun and a rod perpendicular to the ground at high noon during one of the two equinoxes. On or around Sep. 22, 2004, a team of seven students at the Sultan Saiful Rijal Technical College in Bandar Seri Begawan, Brunei measured this angle to be 4.8°. At around the same date, two students from East Waterloo High School in Waterloo, Iowa, measured the angle to be 42°. The students calculated this angle from the length of the rod and the length of the shadow that the rod cast.

Strategy

In this lab you will measure the circumference of Earth by using the above data. You will assume Earth is perfectly round and that the Sun's rays are parallel to each other. The diagram below shows why the angle made by the Sun's rays is less at Brunei, which is slightly north of the equator, than the rays in Iowa.

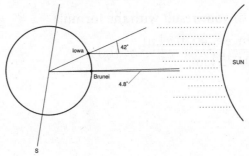

Materials

scissors
protractor
pencil with a fine point
compass

Procedure

1. With scissors, the pencil, and the protractor, create two paper right triangles with the angles for Brunei and Iowa. One leg of the proctrator is proportional to the length of the middle leg of the protractor and the other leg of the protractor is proportional to the length of the shadow. Label each leg of the protractor.

2. Using the compass, draw a large circle on a separate piece of paper. Draw a line through the center of a circle to represent Earth's equator.

3. Place the triangle for Brunei on a quadrant of the circle. Place it so that the protractor

leg representing the shadow is tangent to the circle and the other leg of the protractor is perpendicular to the circle. Orient the triangle so that the shadow falls on the ground in the same direction it fell on the ground when the measurement was taken.

4. Place the triangle for Iowa in the same quadrant in the same way. Shift it until the hypotenuses of both triangles are parallel.

5. Using the protractor, measure the acute angle formed by the two protractor legs representing the middle leg of the protractor.

6. Using an atlas or a globe, measure the distance between the two schools.

Laboratory Activity 1 (continued)

Data and Observations

Angle Between the Protractor Legs	Distance Between the Two Schools

Calculations

The circumference of Earth is given by the formula:

$$\text{circumference} = \frac{\text{distance} \times 370°}{\text{angle}}$$

Analysis

Calculate the percentage error of your result with the formula

$$\text{percentage error} = \frac{\text{circumference} - 40{,}075 \text{ km}}{40{,}075 \text{ km}} \times 100\%$$

Laboratory Activity 1 (continued)

Questions and Conclusions

1. What are the possible sources of error in your calculation?

   _____

   _____

   _____

2. Could the students from Iowa and Brunei calculate the circumference of Earth with only one angle?

   _____

   _____

   _____

3. Using the distance from Iowa to the equator and Brunei to the equator and the above formula, calculate the circumference of Earth.

4. Suppose you mistakenly assumed that Brunei was in the Southern Hemisphere. How would that affect the calculations?

   _____

   _____

   _____

Strategy Check

_____ Do you understand why the Sun casts shadows of different lengths in different parts of the world?

_____ Do you understand how to measure the circumference of Earth?

Copyright © Glencoe/McGraw-Hill, a division of The McGraw-Hill Companies, Inc.

Measuring the Forces that Cause Tides

Chapter 29

Tides on Earth are caused by the gravitational pull of the Moon and the Sun on Earth's oceans. The Moon and Sun cause two bulges each. One bulge comes from the decrease in the gravitational force with distance and the other bulge comes from the centrifugal force acting on the oceans because of the orbital motion of Earth. (We don't ordinarily think of Earth as orbiting around the Moon, but in fact it wobbles in a small circle as the Moon rotates around it.)

Strategy

When an object of moves in a circular path or orbit, a centrifugal force acts upon it. This force is inversely proportional to the square of the time it takes the object to make one revolution (T) and directly proportional the radius of the circular orbit (r). This formula can be written

$$\text{Centrifugal force} = r/T^2$$

The mass of the Moon and Sun can be thought of as being concentrated on a point in space. These two astronomical objects produce a force of gravity that can be represented by drawing a large number of "lines of force" emanating from the single point. The more lines of force per unit area at a particular point the greater the gravitational force at that point. This is shown in the diagram below:

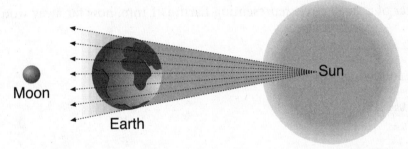

Since the lines of force diverge, there are more lines per unit area on the side of Earth near the Sun or Moon than the side of Earth distant from the Sun or Moon. The diagram shows seven lines of force spread out over a distance d_n on the near side and distance d_f on the far side. The ratio d_f^2/d_n^2 is the same as the ratio of the force of gravity at the near point to the force of gravity at the far point.

Materials

sheets of white paper (5-6)
roll of transparent tape
meterstick
sharp pencil
calculator

Procedure

1. Tape four or five sheets of paper together to make a large sheet of paper.
2. Using the actual values of the radius of Earth and the distance of the Moon from Earth, draw a diagram to scale that shows Earth and the Moon. Draw the Moon as a point source.
3. Using the meterstick, draw two lines from the Moon that are tangent to both sides of Earth.
4. Measure the distance between the two lines at the near side of Earth and at the far side.

The Earth-Moon-Sun System   223

Laboratory Activity 2 (continued)

Data and Observations

Distance of Earth to Moon (r_{Moon})	
Distance of Earth to Sun (r_{Sun})	
Radius of Earth (r_{Earth})	
d_n	
d_f	
Period of rotation of Earth about the Sun (T_{Sun})	
Period of rotation of Earth about the Moon (T_{Moon})	

Calculations

1. Calculate d_f^2/d_n^2 _____.

2. Calculate the ratio r_{Earth}/r_{Sun}_____.

3. If the diameter of a small circle representing Earth is 1 mm, how far away would the Sun

   be_____?

4. Calculate the following expression:

   $$\frac{r_{Sun}}{r_{Moon}} \times \left(\frac{T_{Moon}}{T_{Sun}} \right)$$

Laboratory Activity 2 (continued)

Questions and Conclusions

1. How much stronger is the force of gravity of the Moon on Earth's surface near the Moon compared to the force of gravity on the surface far from the Moon?

_____

_____

2. Why does the decrease in the force of gravity from the Moon cause a bulge in Earth's oceans? On which side of Earth is the bulge?

_____

_____

3. How would you describe the force of gravity of the Sun on Earth's surface near the Sun compared to the force of gravity on the surface far from the Sun?

_____

_____

4. Which force has the greater effect on tides, the centrifugal force from the Moon or from the Sun? By how much?

_____

_____

5. Not counting the centrifugal force, that is, only considering the fact that the force of gravity decreases with distance, which has the greater effect on tides, the Sun or the Moon? By how much?

_____

_____

Strategy Check

_____ Do you understand why the gravitational force decreases with distance?

_____ Do you understand what causes tides?

Making Craters

The surfaces of many planets and moons, including Mercury, Mars, and Earth, have crater indentations indicating that meteorites have crashed there. One meteorite landed near Winslow, Arizona, about 50,000 years ago. In addition to trying to estimate when a crater was formed, scientists attempt to figure out the size of the meteor that struck and from where it might have come. Knowing how surface materials shift on impact during the formation of craters can help scientists in this quest. In this lab, you will observe how craters are made using various "meteorites" that arrive from various distances and from various orbits.

Strategy

You will investigate how size, distance, speed, and angle of impact might affect the formation of craters on the surface of planets and moons.

You will relate how the formation of craters gives scientists clues about the age of craters and where the meteorites originated.

Materials

shallow basin at least 30 cm² (1 ft²),
approximately the size of a cat litter box
one to two bags of unbleached flour per basin
box of instant cocoa powder
several pebbles of various sizes, 1 cm to 4 cm
newspaper or other floor protection
ruler

Safety Precautions 🥽 👕

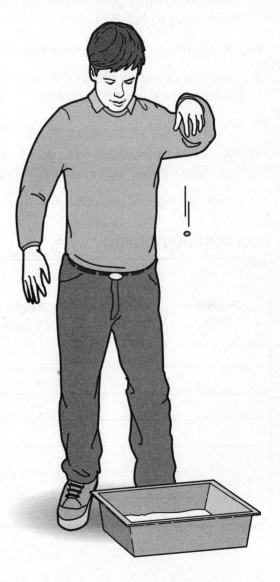

Laboratory Activity 1 (continued)

Procedure

1. Fill a basin with flour about 3–4 cm deep.

2. Sprinkle a little bit of cocoa on the surface to make it easier to observe any splatter the pebbles make in the surface.

3. Observe and record any conditions in the room that might affect your data, such as a running fan that moves the air or vibrates the basin.

4. Drop one of the pebbles into the basin from about eye level.

5. Make initial observations; measure the results with a ruler. Record the results using an illustration, descriptions, and a data chart.

6. Very carefully remove the pebble without disturbing the results and make more observations and measurements.

7. Drop several other pebbles from the same height onto a different spot in the basin, and observe and record the results.

8. Smooth the surface of the flour and sprinkle more cocoa on top.

9. Select one of the pebbles and drop it from various heights, observing the results and recording the data after each drop.

10. Smooth out the flour again and sprinkle more cocoa on top.

11. Toss one of the medium-sized pebbles into the basin with moderate force—first vertically, then from various angles into the basin, being sure to make the toss from the same distance.

12. Record the observations and data after each toss.

Data and Observations

Pebble Size	Height of Drop	Angle of Toss	Observations (crater diameter, circumference, shape, depth, rim uniformity, splatter, etc.)	Room Conditions (air currents, floor or basin vibrations, etc.)

Laboratory Activity 1 (continued)

Questions and Conclusions

1. Do your illustrations show the craters that you expected would be formed? Why or why not?

   _____

   _____

2. Do any of your observations indicate that various sized pebbles, distances from which they were dropped, or various angles at which they were tossed make any difference in the way craters are formed? Explain your answer.

   _____

   _____

3. What does the angle of the pebble toss represent?

   _____

   _____

4. What events might alter the features of a crater after it is formed that could affect scientific observations of the crater?

   _____

   _____

5. Why don't we see more evidence of craters on Earth?

   _____

   _____

6. How do you think Galileo realized the Moon had craters when he first observed the Moon through his telescope?

   _____

   _____

Strategy Check

_____ Can you relate how the characteristics of craters might provide clues to scientists about how and when a crater was formed?

Making a Comet

People have been mystified by comets for centuries. Ancient people thought comets were omens for either good or evil and associated them with major Earthly events. Aside from myth, comets are associated with some amazing statistics. Astronomers hypothesize that comets are composed of rock, frozen gases, and dust from the original nebula that formed the Sun around 4.6 billion years ago. While the core, or nucleus, of a comet is usually a few kilometers across, the cloud that surrounds the nucleus, called the coma, can be 100,000 km across and the tails that stream behind a comet, tens of millions of kilometers long. This length is 1,000 times the diameter of Earth. A comet that follows an orbit that extends to the *Oort Cloud* on the outer reaches of our solar system, 30 trillion km from the Sun, can take 100,000 Earth years to complete one rotation of its orbit. Comets with orbits within the region of the *Kuiper Belt*, close to the orbit of Neptune, take up to 200 Earth years to complete one rotation. Astronomers estimate that there are one trillion comets in our solar system. In this lab, you will make one more.

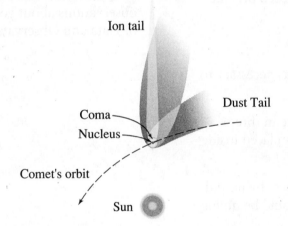

Strategy

You will investigate how comets are formed and what materials make them.

You will relate how the laboratory materials compare to actual comet materials.

Materials

dry ice (2 cups; about 473 mL)
ice chest
newspaper
wooden or rubber mallet
well-insulated work gloves
paper cups
ammonia (2 Tbsp; about 30 mL)
dark corn syrup
    (1/2 tsp; about 2.5 mL)
water (2 cups; about 473 mL)
dirt or sand (1/4 cup; about 59 mL)

mixing bowl, 4-qt (about 3.8 L) or larger,
    plastic or wood (not metal)
strong wood or plastic mixing spoon
    (not metal)
large heavy-weight trash bags with
    no holes (5)
heat lamp
paper plates
paper towels
large cleanup bucket

Laboratory Activity 2 (continued)

Safety Precautions

Procedure

Part A—Making the Comet

1. Read all directions before beginning.

2. Arrange all materials on your work area, premeasured in paper cups.

3. Cut open one garbage bag and line the bowl.

4. Add water to the bowl.

5. Add dirt and stir well.

6. Add corn syrup or other chosen organic material.

7. Add ammonia.

8. Stir and adjust ingredients as necessary to create a muddy sludge.

9. Wearing gloves, place dry ice in three garbage bags that have been placed inside of each other.

10. Crush dry ice into small pieces by pounding with hammer. There should be no big lumps.

11. Add dry ice to the bowl with the other ingredients and stir vigorously.

12. Continue stirring until your comet is almost completely frozen.

13. Remove the spoon and let the comet sit for a few minutes.

14. Lift your comet out of the bowl by the plastic liner and shape it like a snow ball.

15. As soon as the comet is frozen enough to hold its shape, unwrap it.

16. Observe and record the features of your comet in the Data and Observations section.

Part B—Creating a Coma

17. Set up the heat lamp, aiming the light toward the ceiling.

18. With gloves on, hold the comet over the light.

19. Observe and record the effects.

20. Set the comet on a paper plate for display.

21. To clean up, place comets in a large bucket until completely sublimated. Dispose of melted comets in a well-lined garbage can.

22. List the components of actual comets, the ingredients used in this experiment that simulated each component, and any observations about your comet in **Table 1** of Data and Observations.

Laboratory Activity 2 (continued)

Data and Observations

Table 1

Actual Comet Component	Laboratory Comet Component	Observations

Questions and Conclusions

1. What can you infer from this experiment about the materials needed to form actual comets?

   _____

2. What can you infer from this experiment about the processes required to form actual comets?

   _____

3. From where did the basic materials for this lab originally come?

   _____

4. What were the effects of holding the comet over the heat lamp?

   _____

5. What parts of a comet did this experiment simulate and not simulate?

   _____

6. Describe how a comet's coma and tail are formed.

   _____

7. Why do you think astronomer Fred Whipple called comets "dirty snowballs"?

   _____

Strategy Check

_____ Can you describe how comets in our solar system are formed?

_____ Can you list what materials make up comets in our solar system?

How to Measure the Distance to a Star

Chapter 31

Below are replications of two photographs of the same section of the sky taken six months apart. As you can see, one of the stars appears to have moved, while the others have not. The star that moved was much closer to Earth than the other stars, so the motion of Earth around the Sun caused the apparent position of the star to shift.

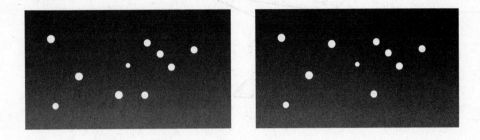

This shift in the position of the star is called the stellar parallax. The figure below shows why the position of the star appeared to shift when viewed six months later.

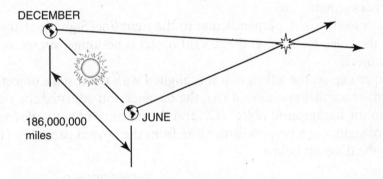

Strategy

In this lab, you will make a device for measuring the distance of a distant object on the ground by measuring its parallax.

Materials

metersticks (2)
paper clips (3)
rubber bands

Laboratory Activity 1 (continued)

Procedure

1. Place a paper clip on each end of a meterstick so that they form a line of sight. This can be done by bending a prong of the paper clip so that it is perpendicular to the meterstick.

2. Fasten the second meterstick so that it forms a cross with the first. Place the third paper clip on the left-hand side of the cross. As shown in the diagram below, you have constructed a device with two lines of sight, one along the length of the ruler and the other at an angle.

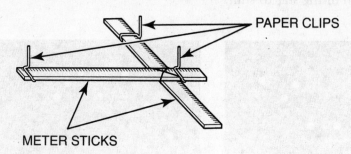

PAPER CLIPS

METER STICKS

3. Select a faraway object whose distance you want to measure. Also take note of an object in the background that is far from the selected object.

4. Align the distant object along the length of the ruler with the two paperclip sights and the object in the background.

5. Take a few steps to the right, perpendicular to the sight line. Sight the background object along the length of the ruler. Notice that the distant object is no longer on the same sight line as the background object.

6. Adjust the paper clip on the left so that it is aligned with the distant object.

7. Measure the distance that you moved (A), the distance you adjusted the paper clip to the new line of sight to the background object (C), and the distance of your eye from the intersection of the new line of sight and a perpendicular line from the moved paperclip (B). These distances are shown in the diagram below.

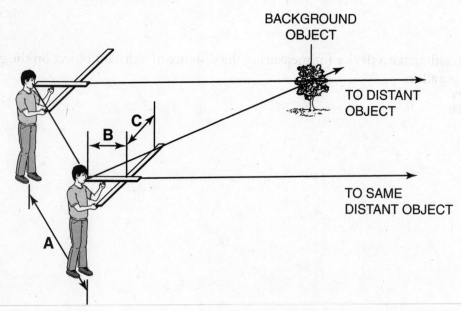

BACKGROUND
OBJECT

TO DISTANT
OBJECT

TO SAME
DISTANT OBJECT

B

C

A

Copyright © Glencoe/McGraw-Hill, a division of The McGraw-Hill Companies, Inc.

Laboratory Activity 1 (continued)

Data and Observations

	Distance
A	
B	
C	

Calculations

The distance to the object is given by the formula

$$\text{Distance of object} = \frac{A \times B}{C}$$

Using this formula, calculate the distance to the object you have selected.

Analysis

Find some other way to measure the distance to the object (for example, by counting the number of paces to the object). Calculate the percentage difference between the two results with the formula

$$\text{Percentage difference} = \frac{\text{parallax method} - \text{direct method}}{\text{direct method}} \times 100\%$$

Laboratory Activity 1 (continued)

Questions and Conclusions

1. Identify the star that moved in the two representations of the same section of the night sky.

_____

_____

2. Why did that star move and not the other stars?

_____

_____

3. The closest star to Earth is Alpha Centauri. The parallax is measured over a six month interval. What distance corresponds to the distance A in the figure on the previous page?

_____

_____

4. When doing this measurement on Alpha Centauri, the distance C is approximately 0.07 mm when B = 10 m. What is the distance from Earth to Alpha Centauri? Show your calculations.

Strategy Check

_____ Do you know what parallax is?

_____ Do you understand how the distance of stars is measured?

Measuring the Diameter of the Sun

Chapter 31

It has been known since ancient times that pinholes can create images just like lenses. The images created by pinholes are much fainter than the images created by lenses because less light goes through the pinhole than through a lens.

Strategy

You will use a pinhole viewer to create an image of the Sun. From the size of the image, you will be able to calculate the Sun's diameter.

WARNING: *Do not use the pinhole to look directly at the Sun.*

Materials

sheet of cardboard
sheet of white paper
aluminum foil (3 cm $\times$ 3 cm)
pin
tape
scissors
candle
meterstick

Procedure

1. Cut a 2 cm $\times$ 2 cm square out of the center of the cardboard.
2. Place the piece of aluminum foil over the opening and tape it in place at the edges.
3. Create a pinhole viewer by puncturing the foil to produce a small hole.
4. Place a candle about 10 cm from the face of the pinhole viewer and place the white sheet of paper a few centimeters on the other side.
5. Test the pinhole viewer by lighting the candle and darkening the room as shown in the diagram below.

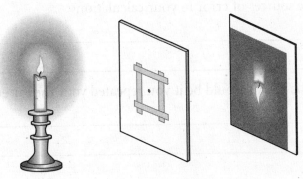

6. Hold the pinhole viewer so that the light from the Sun passes through the hole and falls on the sheet of white paper held behind the hole.
7. Measure the diameter of the image of the Sun (d_i) and the distance of the pinhole to the paper (s_i). Try to make the distance between the pinhole and the paper as large as possible.

Laboratory Activity 2 (continued)

Data and Observations

Diameter of image (d_i)	Distance of pinhole to paper (s_i)

Calculations

The distance of Earth to the Sun is 150,000,000 kilometers. Use the following formula to calculate the diameter of the Sun:

$$\text{diameter of the Sun} = 150{,}000{,}000 \text{ km} \times \frac{d_i}{s_i}$$

The accepted value of the diameter of the Sun is 1,400,000 km. Calculate the percentage difference between your result and the accepted value.

Questions and Conclusions

1. What are the possible sources of error in your calculation?

_____

_____

2. What do you think the results would be if you repeated your measurements?

_____

_____

Strategy Check

_____ Do you know how to make a pinhole viewer?

_____ Do you know how to measure the diameter of the Sun?